my **revision** notes

GET BETTER RESULTS FOR AQA

AQA GCSE
PHYSICS

For A* to C

Steve Witney

D1439559

PHILIP ALLAN
UPDATES

Philip Allan Updates, an imprint of Hodder Education, an Hachette UK company, Market Place, Deddington, Oxfordshire OX15 0SE

Orders

Bookpoint Ltd, 130 Milton Park, Abingdon, Oxfordshire OX14 4SB

tel: 01235 827827

fax: 01235 400401

e-mail: education@bookpoint.co.uk

Lines are open 9.00 a.m.–5.00 p.m., Monday to Saturday, with a 24-hour message answering service. You can also order through Philip Allan Updates website: www.philipallan.co.uk

© Steve Witney 2011

ISBN 978-1-4441-2084-4

Impression number 5 4 3 2 1

Year 2016 2015 2014 2013 2012 2011

Cover photo reproduced by permission of Sashkin/Fotolia

Other photos are reproduced by permission of the following: **p11** *l* Christian Musat/Fotolia, *r* modestlife/ Fotolia; **p43** endostock/Fotolia; **p69** uwimages/Fotolia; **p71** Monkey Business/Fotolia; **p81** braverabbit/Fotolia

Printed in Spain

Hachette UK's policy is to use papers that are natural, renewable and recyclable products and made from wood grown in sustainable forests. The logging and manufacturing processes are expected to conform to the environmental regulations of the country of origin.

P01872

Get the most from this book

This book will help you revise units Physics 1–3 of the new AQA specification. You can use the contents list on pages 2 and 3 to plan your revision, topic by topic. Tick each box when you have:

1 revised and understood a topic

2 tested yourself

3 checked your answers and practised exam questions online

You can also keep track of your revision by ticking off each topic heading through the book. You may find it helpful to add your own notes as you work through each topic.

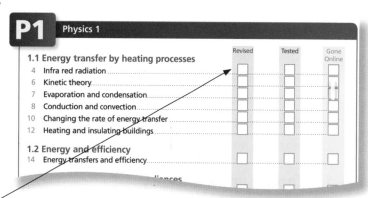

Tick to track your progress

examiner tips

Throughout the book there are exam tips that explain how you can boost your final grade.

Higher tier

Some parts of the AQA specification are tested only on higher-tier exam papers. These sections are highlighted using a solid yellow strip down the side of the page.

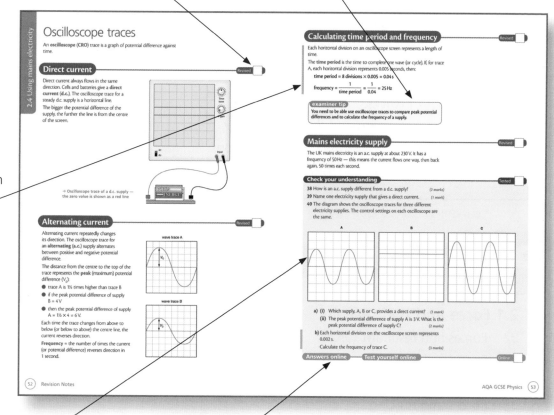

Check your understanding

Use these questions at the end of each section to make sure that you have understood every topic.

Go online

Go online to check your answers at **www.therevisionbutton.co.uk/myrevisionnotes**.

Here you can also find extra exam questions for topics as well as podcasts to support you when getting ready for the big day.

Contents and revision planner

P1 Physics 1

P2 Physics 2

P3 Physics 3

Infra red radiation

Infra red and thermal radiation

Revised ▢

When an object **absorbs** any type of electromagnetic radiation it gets hotter.

● Absorbing infra red radiation gives the most obvious heating effect.

● **Infra red radiation** is often called thermal radiation or heat — they are the same thing.

In the diagram, the hot element of the electric fire emits, or radiates, a lot of infra red radiation every second.

● The energy is transferred through the air by electromagnetic waves.

● The person in front of the fire feels warm because she absorbs some of this energy.

● Objects heat up (increase in temperature) when they absorb infra red radiation faster than they emit it.

● The hotter an object, the more infra red radiation it radiates (transfers) each second.

All objects absorb and emit infra red radiation

Revised ▢

But some surfaces are better at it than others. The chart shows the difference between surfaces.

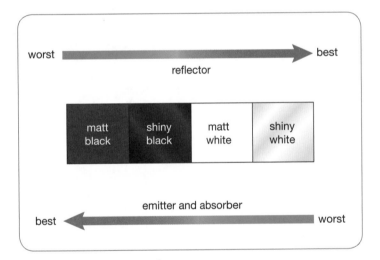

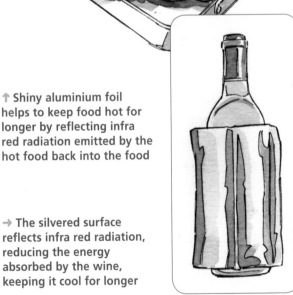

● Dark-coloured, matt surfaces absorb and emit infra red radiation faster than light-coloured, shiny surfaces.

● Good absorbers of infra red radiation are also good emitters of infra red radiation.

● Light-coloured, shiny surfaces are good reflectors of infra red radiation.

● Shiny surfaces reduce energy transfer by infra red radiation.

↑ **Shiny aluminium foil helps to keep food hot for longer by reflecting infra red radiation emitted by the hot food back into the food**

→ **The silvered surface reflects infra red radiation, reducing the energy absorbed by the wine, keeping it cool for longer**

Comparing surfaces

The diagram of the two different containers shows a simple way of comparing the effect of the type of surface on the rate of absorption.

● Repeating the experiment would always give the same result — this makes the measurements both **repeatable** and **reproducible**.

● An **error** in reading the thermometer could give an **anomalous** result — one that does not fit the pattern.

● Using a temperature sensor and data logger would increase the **accuracy** of the temperature measurements — this is because you are less likely to misread the value (giving a **random error**).

● Increasing the accuracy means that the recorded temperatures are closer to the **true temperature**.

● A temperature sensor has a greater **resolution** — it responds to smaller changes than a thermometer that only shows readings to the nearest 0.5 °C.

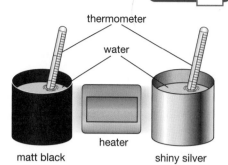

↑ The temperature inside the black container goes up more quickly than the temperature inside the shiny silver container

examiner tip
When the examiner asks you to 'explain', you should apply some logical thinking — for example, give the reasons for something happening in terms of theory.

Check your understanding

1 Explain why a hot pie placed in a refrigerator cools down.

(2 marks)

2 A hollow metal cube is filled with boiling water. Each side of the cube has a different colour or texture. The values in the table are taken with an infra red radiation detector facing each side of the cube. The higher the value, the greater the amount of infra red radiation emitted by the surface.

Side	A	B	C	D
Detector value	0.37	0.72	0.76	0.26

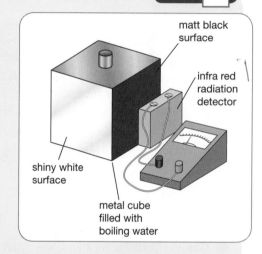

a) Match each side (A–D) to one of the colours and textures in the list below. *(4 marks)*

matt black	matt white	shiny black	shiny silver

examiner tip
You need to extrapolate the graph. This means continue drawing the line following the same pattern.

b) Why is it important in this experiment that the detector and meter could measure small changes?

(1 mark)

3 The graph shows how the temperature inside a room changes.

a) At which point (X, Y or Z) is the room absorbing and emitting infra red radiation at the same rate? Give a reason for your answer. *(2 marks)*

b) The temperature of the room continues to increase at the same rate for a further hour. What is the room temperature at 13.00 h?

(2 marks)

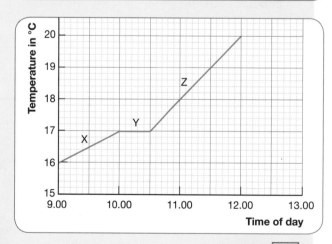

Kinetic theory

- A substance can exist as a solid, a liquid or a gas. These are the **three states of matter**. Ice, water and steam are the same substance, but in different states of matter.

- **Kinetic theory** states that all matter is made up of small **particles** that are constantly moving — the higher the temperature, the faster the particles move.

- When close together, the particles attract each other strongly.

- Kinetic theory is a **model** used by scientists to explain how solids, liquids and gases behave.

- **Solid**: the particles are packed close together in a fixed pattern. The particles are constantly vibrating but held in position by strong attractive forces. If a solid is heated, the particles vibrate faster and take up more space, which means the solid has expanded a little.

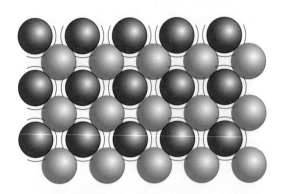

- **Liquid**: the particles are close together, but the forces are not strong enough to hold them in a fixed pattern. They are able to move and slide about. This means that a liquid can flow and changes its shape to fit any container. Particles at the surface may have enough energy to break away and become a gas (or vapour).

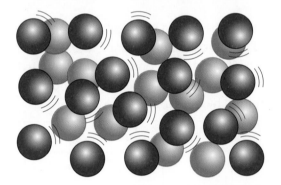

- **Gas**: the particles are a long way apart with virtually no forces between them. This is why a gas is easy to compress. The particles have a rapid, random motion and quickly fill any space available to them.

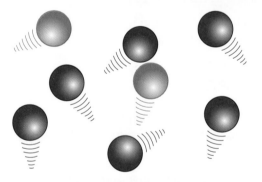

Check your understanding
Tested

4 Use kinetic theory to explain why a gas can be compressed much more than a liquid.

(1 mark)

5 Compare the forces between particles in a solid with the forces between particles in a liquid.

(1 mark)

6 Use kinetic theory to explain why a liquid can be poured.　*(1 mark)*

7 Describe what is meant by 'rapid random motion'.　*(2 marks)*

Answers online — **Test yourself online**
Online

Evaporation and condensation

Evaporation

Revised

Evaporation happens when a liquid changes into a gas without boiling.

Evaporation can be explained using kinetic theory. The particles in a liquid have a range of speeds, with some particles moving much faster than the others. If one of these faster particles is close to the liquid surface it might have enough kinetic energy to break away and escape from the liquid.

Liquids can be made to evaporate faster by:

- Increasing the temperature — increases the speed of the particles, so more will have enough energy to escape.
- Increasing the surface area — lets more of the fastest particles be next to the surface, so particles escape at a faster rate.
- Having a draught or wind — this will move many of the escaping particles away from the surface before they fall back in.
- Reducing the humidity — reduces the amount of water vapour in the air, reducing the number of particles from the water vapour falling back into the liquid.

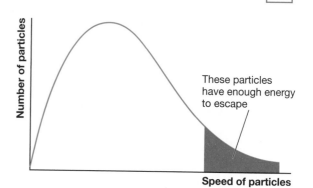

↑ **The particles with the highest speeds have enough energy to evaporate**

Evaporation causes cooling

Revised

- When a liquid evaporates, it loses its fastest particles — the ones with the most energy.
- The average energy of the particles left behind is less than before.
- The lower the average energy, the lower the temperature.

examiner tip

In the exam, always explain evaporation by using the kinetic model (particles).

Condensation

Revised

Condensation happens when a gas changes back into a liquid.

When warm, damp air hits a cold surface, it cools and some of the water vapour in the air changes back into a liquid — this is condensation.

Check your understanding

Tested

8 Explain, using kinetic theory, why blowing across the top of a hot cup of tea cools the tea more quickly.
(2 marks)

9 If you get out of a swimming pool and do not dry yourself, you will soon begin to feel cold. Explain why.
(2 marks)

10 Explain why boiling water in a saucepan often leads to 'condensation' on the windows.
(2 marks)

Answers online   Test yourself online   Online

Conduction and convection

The transfer of energy by **conduction** and **convection** involves the movement of particles.

> **examiner tip**
> Conduction and convection are similar words, but different processes — make sure you know which an exam question is asking about.

Conduction happens mainly in solids
Revised

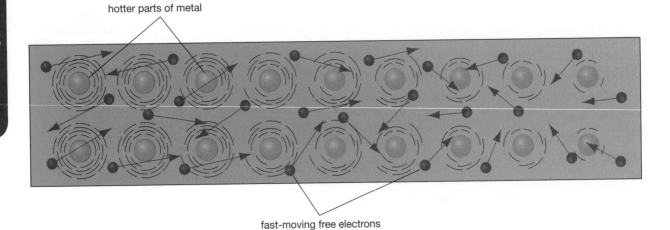

atoms vibrate faster in hotter parts of metal

fast-moving free electrons

- At the hot end of the conductor, energy is transferred to the atoms increasing their kinetic **energy**.
- The atoms **vibrate** faster and with a bigger amplitude.
- These atoms collide with neighbouring atoms.
- The extra kinetic energy is passed from atom to atom.
- This is a slow process.

Metals are better conductors than non-metals
Revised

- Metals have **electrons** that are free to move.
- These free electrons transfer energy rapidly.
- Non-metals have no free electrons, so conduct energy slowly.
- A poor conductor is a good **insulator**.

Air is a good insulator
Revised

← A polar bear has a thick fur coat. The fur traps small pockets of air. The air reduces the energy transferred by conduction, helping to keep the polar bear warm

Convection happens only in liquids and gases

Revised

Convection is the transfer of thermal energy by the movement of a liquid or gas due to differences in **density**.

● Air particles gain energy and move around more quickly.

● The particles move apart, taking up more space.

● The warm air **expands**, becoming **less dense** than the colder air above.

● The warmer air rises, pushing colder, denser air downwards.

> **examiner tip**
>
> To explain how water is heated by convection, just change the word 'air' to 'water'. There is no need to learn a whole new explanation.

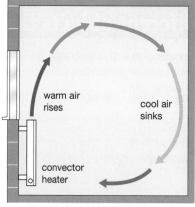

⬆ The heater creates an air flow called a convection current

A hot-water system uses convection

Revised

● A boiler heats the water.

● Hot, less dense water rises to the hot-water tank.

● Colder, denser water falls from the cold-water tank.

● A convection current circulates the water.

Question: 'When is a radiator not a radiator?'

Answer: 'When it's a radiator in a central heating system.'

The radiators in a central heating system are really convector heaters. By heating up the air, they create a convection current that transfers the energy around the room.

To transfer a lot of energy by radiation, they would need to be a lot hotter than they are.

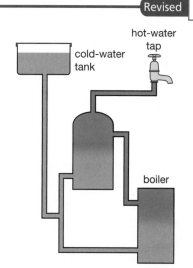

Check your understanding

Tested

11 Greg and Jill pack an insulated picnic box with food. Greg wants to put the frozen ice pack at the bottom of the bag. Jill says it would be better if the ice pack were on top of the food at the top of the bag.

Where in the box should the ice pack go? Explain the reason for your answer. *(3 marks)*

12 Which two statements describe the transfer of energy by convection through water? *(2 marks)*

> **examiner tip**
>
> Write out the two correct statements — it will help you to remember them.

 A   Energy is transferred by electromagnetic waves.

 B   Energy transfer does not involve particles.

 C   Energy is transferred because the water expands.

 D   Energy is transferred by free electrons.

 E   Energy is transferred because the density of water changes as it is heated.

13 Explain why copper is a better thermal conductor than plastic.

(1 mark)

Answers online   Test yourself online   Online

Changing the rate of energy transfer

Different objects transfer energy at different rates Revised

The **rate** of transfer of energy by heating, to or from an object, depends on:

- the material the object is made from
- the surface area and volume of the object
- the surface that the object rests on
- the difference in temperature between the object and its surroundings

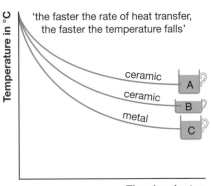

'the faster the rate of heat transfer, the faster the temperature falls'

ceramic — A
ceramic — B
metal — C

Temperature in °C

Time in minutes

Comparing graph lines	
A, B or C	The bigger the temperature difference between the water and the air, the faster the rate of energy transfer
A and C	A metal mug conducts heat away from the water more quickly than a ceramic mug
A and B	Increasing the surface area increases the rate of heat transfer

To make the comparison of these graphs a fair test:

- only the shape or material of the mug was changed (the **independent variable**)
- the volume and starting temperature of the water was kept the same (**control variables**)

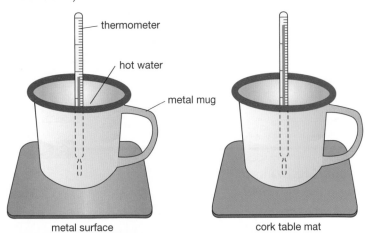

thermometer

hot water

metal mug

metal surface

cork table mat

← A metal surface conducts energy away from the water in the mug faster than a cork surface. The cork is a good insulator. Energy transfer to or from an object can be slowed by covering the object with an insulating material

↑ The large surface area of the cooling fins increases the rate of energy transfer from the engine by convection and radiation

↑ The skier has several layers of clothes on — a good idea in the cold. Air is trapped between the layers as well as in the fabric of the clothes. The trapped air reduces the rate of energy loss by conduction and convection

Animal adaptations

Revised

Animals have been adapted to survive in their environment.

- An Arctic fox has a compact body and small furry ears — this gives it a low surface area-to-volume ratio, which reduces the rate of heat loss.
- A desert fox has a large surface area from which to lose heat — the sandy coloured fur helps reflect infra red radiation.

Vacuum flasks

Revised

Vacuum flasks keep hot drinks hot and cold drinks cold. Energy transfer by conduction, convection and radiation is reduced by:

- the vacuum between the double glass walls (conduction and convection)
- the silvering on the glass walls (radiation)
- the plastic stopper (conduction and convection)

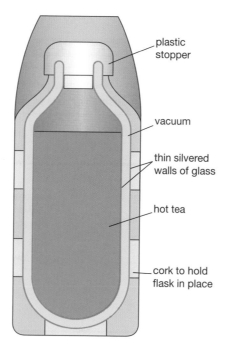

plastic stopper

vacuum

thin silvered walls of glass

hot tea

cork to hold flask in place

Check your understanding

Tested

14 Explain why woollen gloves keep your hands warm. (2 marks)

15 Explain why a rabbit fluffs out its fur on a cold day. (2 marks)

16 What features of a vacuum flask are designed to reduce energy transfer by conduction? (1 mark)

17 A car radiator has many cooling fins and is usually painted black. Explain how these features of a radiator help to cool the car engine. (3 marks)

Answers online **Test yourself online**

Online

Heating and insulating buildings

Keeping our homes warm

Revised

Keeping our homes warm is about reducing the rate of energy transfer to the air outside. Most methods of insulation involve trapping air:

- Air trapped in small pockets cannot move far, so energy transfer by convection is reduced.
- Air is a good insulator, so energy transfer by conduction is reduced.

Double-glazed windows have two sheets of glass with air (or air and argon) mixture in between. Air, argon and glass are good insulators, reducing energy loss by conduction. The gap between the sheets of glass is thin, reducing energy loss by convection. The glass may also have a coating that reflects infra red, reducing heat loss by radiation.

Shiny materials, often put in the loft or behind radiators, are good reflectors to reduce energy transfer by radiation.

Draught excluders help to reduce the effect of convection currents pulling cold air into the house.

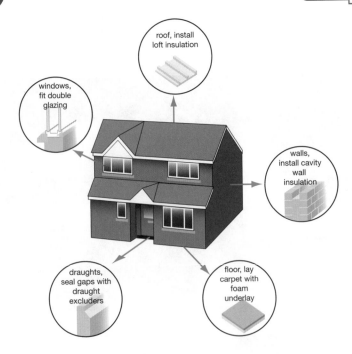

roof, install loft insulation

windows, fit double glazing

walls, install cavity wall insulation

draughts, seal gaps with draught excluders

floor, lay carpet with foam underlay

↑ **Heat transferred through the roof, walls, windows, floor and doors of a house can be reduced by using insulation**

examiner tip

An exam question may not be about house insulation, but if it's about keeping warm then it's almost certainly about trapped air reducing the rate of energy loss by convection and conduction.

U-values

Revised

U-values measure how good a material is as an insulator:

- the lower the U-value, the better the insulation provided by the material
- the higher the U-value, the faster the material will transfer thermal energy.

examiner tip

You don't need to memorise U-values or its unit. If you need to compare materials, U-values will always be given in the question.

Solar panels

Revised

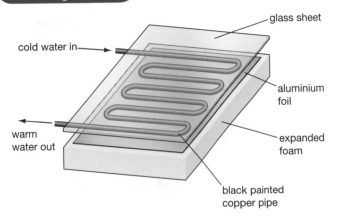

glass sheet

cold water in

aluminium foil

warm water out

expanded foam

black painted copper pipe

←**Solar panels use energy radiated by the Sun to provide hot water**

Specific heat capacity

The specific heat capacity of water is 4200 J/kg°C. This means that 4200 joules of energy are needed to increase the temperature of 1 kilogram of water by 1° Celsius. It also means that when 1 kilogram of water cools by 1° Celsius, it loses 4200 joules of energy.

The high specific heat capacity of water explains why it is so good at storing thermal energy.

The energy needed to increase the temperature of an object is calculated using the equation:

$E = m \times c \times \theta$

energy transferred = mass × specific heat capacity × temperature change

● E in joules, J

● m in kilograms, kg

● c in J/kg°C

● θ in degrees Celsius, °C

This equation can also be used to calculate the energy lost when an object cools down.

examiner tip

You do not need to learn the equations — but you must be able to find the right one from the equation sheet and be able to use it.

Check your understanding

18 The diagram above (under the heading 'Solar panels') shows a model solar panel water heater made by a student. Explain why each of the features labelled in the diagram has been included in the design. *(6 marks)*

19 The *U*-values for different types of double glazing are given in the table. The gap between the glass, which can be 12 mm or 20 mm, is filled with either air or argon.

Data set	Glass type	*U*-value	
		12 mm gap	**20 mm gap**
1	Type A with air	2.9	2.8
2	Type A with argon	2.7	2.6
3	Type B with air	1.9	1.8
4	Type B with argon	1.6	1.5

a) Windows with a 20 mm gap cost more than those with a 12 mm gap. Is the extra expense worthwhile? Give a reason for your answer. *(1 mark)*

b) The data were obtained with a temperature difference of 15°C between one side of the window and the other. How would these figures change if the temperature difference was increased to 20°C? Give a reason for your answer. *(2 marks)*

c) Which sets of data should be compared to decide which type of glass, A or B, makes the most energy-efficient double-glazed window? *(1 mark)*

20 Calculate the energy transferred from a 20 kg block of concrete that cools by 40°C.

The specific heat capacity of concrete is 800 J/kg°C. *(2 marks)*

Answers online — **Test yourself online**

Energy transfers and efficiency

We cannot create energy from nothing

Revised

Energy can be:
- **transferred** (moved) from one place to another
- **stored**
- **dissipated** (spread out)

Remember that energy cannot be created or destroyed.

Energy makes things happen

Revised

Chemical energy stored in fuels is useful, but the fuel must be burned for the energy to make anything happen. As the store of chemical energy goes down, something is heated and its temperature goes up. Burning petrol in a car engine transfers chemical energy to kinetic energy, heat and sound.

A **Sankey diagram** is used to show percentage energy transfers — the wider an arrow, the more energy is transferred. The Sankey diagram for the petrol engine shows that the energy input and the total energy output are the same — the energy has been conserved. But not all output energy is useful — the heat and sound are wasted energy.

A Sankey diagram drawn to scale can be used to calculate the efficiency of an appliance.

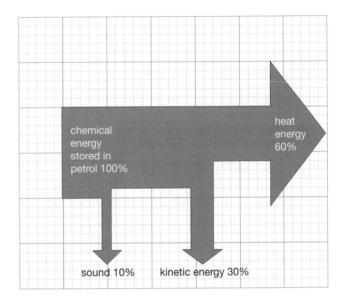

↑ **Sankey diagram for a petrol engine**

> **examiner tip**
> If you sketch a Sankey diagram, there is no need to draw it to scale. But remember to label each arrow.

An **efficient** appliance is good at transferring input energy into useful output energy.

$$\text{efficiency} = \frac{\text{useful energy out}}{\text{total energy in}} \ (\times\ 100\%)$$

- Energy is measured in **joules** (J).
- Efficiency has no unit.
- Efficiency is given as a decimal number (or a percentage).

> **examiner tip**
> To calculate efficiency, always divide by the biggest number because efficiency is always less than 1 (or 100%).

Energy can never vanish

Revised

- All energy, useful and wasted, is eventually transferred to the surroundings, which become warmer.
- The energy is shared between lots of molecules, so it becomes spread out.

- As the energy spreads out, it becomes more difficult to use for further energy transfers.
- The energy has not vanished — it's just not useful any more.

In some devices, energy normally wasted can be used usefully. On cold days the inside of a car is warmed by energy from the engine that is normally wasted.

Payback time and cost-effectiveness

Revised

Payback time is not the same as cost-effectiveness. **Payback time** is the time it takes to recover the money spent on reducing energy consumption from the money saved on energy bills. For example, if £240 is spent on loft insulation and it reduces the annual energy bill by £80 then the payback time is 3 years.

How **cost-effective** it is to reduce energy consumption depends on:

- the initial cost
- the replacement time

Method of reducing energy transfer	Cost to install/£	Annual saving on energy bills/£	Replacement time/years	Total saving over 5 years/£	Total saving over 20 years/£
Draughtproofing	75	25	5	$(5 \times 25) - 75 = 50$	$(20 \times 25) - (4 \times 75) = 200$
Temperature controls on radiators	120	20	20	$(5 \times 20) - 120 = -20$	$(20 \times 20) - 120 = 280$

Draughtproofing has the shorter payback time (3 years), and over 5 years is the most cost-effective. But over 20 years, draughtproofing must be replaced four times, so over 20 years the radiator controls are more cost-effective.

Check your understanding

Tested

21 Use the data in the table to find the number of years it takes for the radiator controls to become more cost-effective than the draughtproofing. *(2 marks)*

22 A fluorescent light transforms 25% of the input energy to light and 75% to heat. Describe what happens to the output energy from the lamp. *(2 marks)*

23 A television is 65% efficient. Each second it is on, the useful energy output is 520 J.

a) Calculate the energy input to the television each second that it is switched on. *(2 marks)*

b) When it is switched to standby, the energy input to the television halves. Why should televisions be switched off, rather than switched to standby? *(1 mark)*

24 A solar cell is 8% efficient. What is the energy output each second when the energy input to the panel is 2400 J each second? *(2 marks)*

Answers online — Test yourself online

Online

Electrical power and energy costs

Electrical energy is easily transferred. This is what makes it so useful, and why so many appliances are designed to work from an electricity supply. Examples are given in the table.

Energy input to appliance	Appliance	Useful energy output from appliance
Electrical	Loudspeaker, buzzer, bell	Sound
Electrical	Grill, toaster, fire, iron, kettle	Heat
Electrical	Lamp, computer screen, digital display	Light
Electrical	Motor	Kinetic
Electrical	Rechargeable battery, phone charger	Chemical

examiner tip

Think about what you use a device for, and the useful energy output should be obvious. You listen to an MP3 player — so the useful output must be sound.

Power

Revised

Power is the **rate at which an appliance transfers energy**. It is measured in **watts** (W):

● 1 watt = 1 joule of energy transferred in one second (1 J/s)
● 1 kilowatt (kW) = 1000 watts (W)

The total electrical energy transferred by an appliance depends on:

● how long the appliance is used
● the power of the appliance

examiner tip

The efficiency of an appliance can be calculated using power — just change the word 'energy' to 'power' in the efficiency equation.

Cost of the energy

Revised

First calculate the energy transferred by an appliance using the equation:

$E = P \times t$

energy transferred = power × time

● E in kilowatt-hours, kWh
● P in kilowatts, kW
● t in hours, h

Then:

total cost = number of kilowatt-hours × cost per kilowatt-hour

This is the only time that we calculate energy in **kilowatt-hours**.

examiner tip

To get the right cost, you must make sure power is in kilowatts and time is in hours. Also remember answers should be realistic!

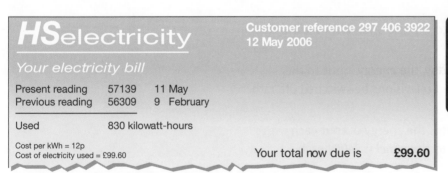

HSelectricity

Customer reference 297 406 3922
12 May 2006

Your electricity bill

| Present reading | 57139 | 11 May |
| Previous reading | 56309 | 9 February |

| Used | 830 kilowatt-hours |

Cost per kWh = 12p
Cost of electricity used = £99.60

Your total now due is   **£99.60**

⬆ The readings on the electricity meter are used to calculate the total energy cost over 3 months

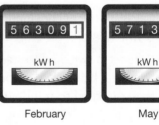

February   May

⬆ An electricity meter records the energy supplied in kilowatt-hours

Choosing the right appliance for the job

Revised

Different appliances that do the same job may have particular advantages and disadvantages. For example, there are many different types of electric heater. Which one do you choose?

For a quick blast of heat in a garage, shed or kitchen, a fan heater would probably be first choice. To keep the chill out of a bedroom, it's probably the oil-filled radiator.

> **examiner tip**
>
> If you are asked to compare different appliances, do not just copy the given information — add something to it.

Heater type	Advantages	Disadvantages
2.5 kW fan heater	● Two power settings ● Rapidly warms a room ● Small size	● Expensive to run on full power for long ● Fan can be noisy ● Does not give same room temperature throughout
400 W oil-filled radiator	● Low operating cost ● Makes no noise ● Maintains a steady room temperature	● One power setting ● Takes time to warm up a room ● Expensive compared with other heaters

Check your understanding

Tested

25 Calculate the cost of using a 1600 W hair dryer for 20 minutes every day, Monday to Saturday. One kilowatt-hour of energy costs 15p.

(3 marks)

26 How much energy does an 11 W lamp switched on for 12 hours transfer? *(2 marks)*

27 A household electricity meter reads 35 276 at 9 a.m. By 9 p.m. on the same day, the following appliances have been used:

a 3 kW oven for 2.5 hours

a 1.5 kW dishwasher for 1 hour

a 2 kW tumble dryer for 30 minutes

What is the reading on the electricity meter at 9 p.m? *(3 marks)*

Answers online   Test yourself online   Online

Generating electricity

Most electricity in the UK is generated in power stations using the energy from a non-renewable fuel. **Non-renewable fuels** take millions of years to replace. Once they are gone, they are gone forever.

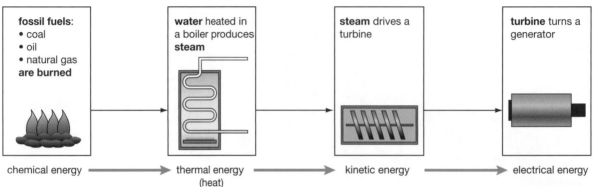

fossil fuels:	**water** heated in	**steam** drives a	**turbine** turns a
• coal	a boiler produces	turbine	generator
• oil	**steam**		
• natural gas			
are burned			

chemical energy ⟶ thermal energy (heat) ⟶ kinetic energy ⟶ electrical energy

Small-scale gas power stations use the heat from burning gas to produce fast-moving air, which drives the turbine directly. Large combined cycle gas turbine (CCGT) power stations use both gas turbines and steam turbines to generate electricity.

Nuclear power stations generate electricity in a similar way to those that burn coal. But the fuel is not burned in a nuclear power station. Steam is produced using the heat given out when the uranium or plutonium fuel atoms undergo **nuclear fission** reactions.

Different types of power station have different **start-up times** — this is how long it takes a power station to begin to generate after it has been closed down.

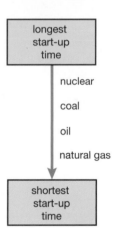

longest start-up time

nuclear

coal

oil

natural gas

shortest start-up time

The National Grid

The **National Grid** is a network of cables and transformers that transfer energy. It links the power stations that generate electricity to the consumers who use it.

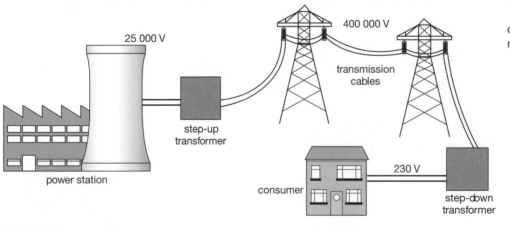

25 000 V

400 000 V

transmission cables

step-up transformer

power station

consumer

230 V

step-down transformer

←The essential parts of the National Grid network

Transferring electrical energy

Transferring electrical energy at a high voltage increases efficiency.

● The power station generates electricity.

- A **step-up transformer** increases the voltage (potential difference) across the transmission cables.
- For a particular power, increasing the voltage decreases the current through the cables.
- A smaller current means less energy is transformed into heat and wasted.
- Reducing the amount of waste energy makes the transfer of electricity more efficient.

At the end of the transmission cables, a series of **step-down transformers** are used to decrease the voltage (potential difference) to a safe value for consumers to use.

examiner tip

You do not need to know how transformers work, only why they are used.

Overhead or underground?

Revised

Most of the National Grid transmission cables can be seen because they go overhead. However, some cannot be seen — these run underground.

examiner tip

You need to remember the advantages and disadvantages of overhead and underground cables.

Advantages and disadvantages of overhead and underground cables

	Overhead	Underground
Advantages	- Relatively cheap to install - Quick and easy to repair - Electrical insulation provided by the air	- No hazard to aircraft - Cannot be seen - Normally no shock hazard - Unlikely to be damaged by the weather
Disadvantages	- Hazard to low-flying aircraft - Spoil the landscape - Risk of fatal electrocution - Can be damaged by severe weather	- Expensive to install - Longer and more expensive to repair - Need layers of electrical insulation and a cooling system

Check your understanding

Tested

28 Why does a small-scale gas-burning power station have a faster start-up time than a coal-burning power station? *(1 mark)*

29

Fuel	Building cost	Fuel cost	Operating cost	Decommissioning cost
Nuclear	4.2	0.4	0.6	2.7
Coal	2.0	0.3	0.7	0.1
Oil	2.2	3.5	1.2	0.1
Natural gas	2.3	1.3	0.3	0.1

The table gives the estimated cost (in pence) of generating 1 kWh of electricity from different fuels. Give the order (most expensive first) for the total cost of generating 1 kWh of electrical energy. *(2 marks)*

30 How is the production of electricity in a nuclear power station different from that in an oil-burning power station? *(1 mark)*

31 It takes 5 years to build a new nuclear power station — each year, the power used in construction is 250 MW. Once operating, the power station generates 1000 MW of power. How long will it be from the start of construction until the power station has generated more electrical energy than the energy used to build it? *(2 marks)*

Answers online — Test yourself online

Online

Renewable energy resources

Renewable energy resources are replaced as quickly as they are used. Most renewable energy resources used to generate electricity do not burn fuels — the energy to drive the turbine comes directly from the renewable resource.

Energy from wind, waves and tides

Revised

Energy from wind, waves and tides is freely available. The kinetic energy of wind and of moving water can be used to drive a turbine.

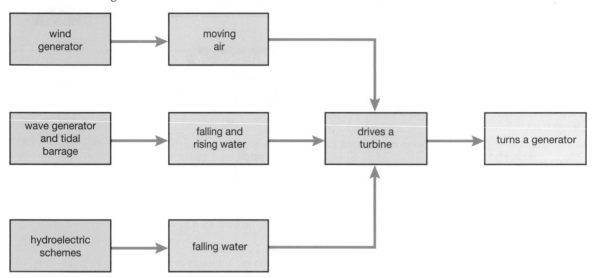

In an oscillating water column generator, the movement of the waves forces the air to move, which then drives the turbine.

Regular tides cause the water level in an estuary to rise and fall. Turbines built into a barrage across the estuary are driven by the moving water. Smaller systems use submerged generators that are driven by underwater currents.

Hydroelectric systems trap water behind a dam. The dam is built across a river to give a continuous supply of water. When released, the water falls, driving the turbines.

> gravitational potential energy →
> kinetic energy → electrical energy

A hydroelectric pumped-storage power station uses surplus electricity to pump the water back up behind the dam, effectively storing energy for later use. This makes the power station ready for the next time there is a peak in demand.

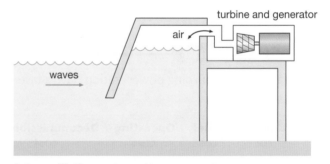

↑ **An oscillating water column generator**

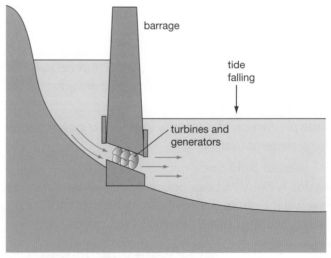

↑ **A generator built into a tidal barrage**

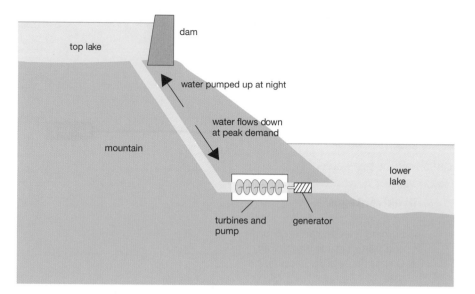

←A pumped storage power
station

Geothermal energy
Revised

In some volcanic areas, the hot rocks below the Earth's surface are used to
heat water, often turning it into steam. The steam that rises to the surface
can be used to drive turbines.

Solar energy
Revised

Solar cells transform the energy of the Sun's radiation directly into
electricity. With a solar thermal tower, solar energy is used to heat water
or oil, the vapour of which is then used to drive a turbine.

Biofuels
Revised

Biofuels come from growing plants. As they grow, they store chemical
energy. Burning a biofuel allows this energy to be used to generate
electricity in the same way as a coal-burning power station. Biofuels
include wood, straw, olives and palm nuts.

Check your understanding
Tested

32 Give one advantage of a tidal barrage compared to a wind
generator.
(1 mark)

33 Which type of power station involves the damming of a river?
(1 mark)

34 What must be done to ensure that wood is a renewable energy
resource?
(1 mark)

35 Use the following information to calculate the area of solar cells
needed to generate 200 kW of electrical power.

Average power from the Sun = 400 W/m^2

Efficiency of solar cells = 12.5%
(3 marks)

examiner tip
Remember to show your working in
calculations. Marks may be available
for each stage of the calculation.

Answers online   **Test yourself online**
Online

Comparing energy resources

The tables below compare the advantages and disadvantages for the fuels and energy sources used to generate electricity.

Non-renewable fuels are reliable

Provided a fuel is available, electricity can always be generated.

Resource	Advantages	Disadvantages
Coal and oil	● Easy to transport ● A concentrated source of energy	● Burning releases carbon dioxide and sulfur dioxide into the air ● Oil has many other important uses
Natural gas	● Quick to start up and to switch off	● Burning releases carbon dioxide into the air
Nuclear fuels	● No polluting gases produced ● A very concentrated source of energy — a small mass of fuel gives a huge amount of energy	● Some radioactive waste must be stored for thousands of years ● Serious accidents may release radiation over large areas ● Expensive to decommission a power station at the end of its useful life

Carbon dioxide emissions could be reduced by the new technology of **carbon capture**. The idea is to store the carbon dioxide before it enters the atmosphere. Natural containers, such as the chambers left in exhausted oil and gas fields under the sea, would be big enough to store the vast amounts involved.

Renewable energy resources

Generally, renewable energy resources:

● produce less chemical pollution than non-renewable fuels

● give free energy — although transferring the energy to electrical energy can be expensive

Resource	Advantage	Disadvantage
Wind	● Running costs are low ● Land around turbines can be used for farming	● Some people think they spoil the landscape (visual pollution) and make unwanted noise ● Unreliable, only generate when the wind is strong enough ● Dilute energy resource, so a lot of turbines are needed
Waves	● Running costs are low	● Need to be very strong to withstand the force of a very rough sea
Tides	● Reliable, the tides happen twice a day, every day	● Barrages are expensive to build ● Barrages destroy the habitat of wading birds and other wildlife

Resource	Advantage	Disadvantage
Hydroelectric	● Can generate large amounts of electricity	● Large areas of land may be flooded ● Flooding may destroy people's homes and affect plant and animal life
Pumped-storage	● Can be switched on quickly to meet peak demands	● Cannot run continuously
Solar	● Ideal for remote places ● Solar thermal towers generate a lot of electricity	● Electricity generation depends on light intensity ● Large areas of solar cells needed to generate a lot of power
Geothermal	● Massive amounts of energy available	● Not always practical to extract the energy from the Earth
Biofuel	● Adds no additional carbon dioxide to the atmosphere	● Large areas of land needed to grow the crop

Increased demand for electricity

Revised

Increased demand for electricity means tough decisions have to be made. The UK government wants to increase the amount of electricity generated from renewable energy resources. This could be done by erecting many more wind farms, both onshore and offshore. But is this the answer? People living close to any proposed development may not want it. Balanced arguments are needed so that an informed decision can be made. Wind turbines may be the answer for people on a remote, wind-swept island, but not for those living in a new town built in a sheltered inland area.

Many governments want to reduce the emissions of carbon dioxide. In the UK, we could generate more electricity from nuclear fuels and less from fossil fuels. This would also make the UK less dependent on imported gas. But how safe are nuclear power stations? And just where do you store the nuclear waste? These issues must be considered before making any decision about the future of nuclear energy.

Check your understanding

Tested

36 Give two reasons to support the building of new nuclear power stations. *(2 marks)*

37 The doctor in a remote African village stores medicines in a freezer. Solar cells are used to generate the electricity needed to power the freezer. Give one reason why solar cells are used. *(1 mark)*

38 Why does a coal-burning power station take up less land than a wind farm that produces the same amount of electricity? *(1 mark)*

39 Explain why it is important to increase the amount of electricity generated using renewable energy resources. *(2 marks)*

examiner tip

Use the information given in an exam question to help you to formulate arguments for or against the use of a specific energy resource.

Answers online   Test yourself online

Online

Properties of waves

All waves move energy

Revised

All waves move energy from one place to another. For example, throwing a stone into a pond causes ripples to spread out. As the water particles vibrate up and down, energy is transferred outwards from where the stone hit the water.

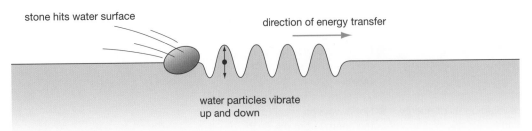

stone hits water surface

direction of energy transfer

water particles vibrate up and down

Transverse waves

Revised

● The **oscillation** (vibration) creating the wave is at right angles to the direction of energy transfer.

direction of energy transfer

oscillation

direction of energy transfer

movements of hand from side to side

the tape moves from side to side

←Moving a stretched 'slinky' spring from side-to-side sends a transverse wave through the spring

● **Light** is an example of a transverse wave.

Longitudinal waves

Revised

● The oscillation (vibration) creating the wave is parallel to the direction of energy transfer.

oscillation

direction of energy transfer

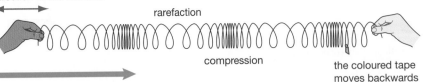

direction of the vibration

rarefaction

this end is held still

compression

the coloured tape moves backwards and forwards

direction of energy transfer

←Pushing a stretched 'slinky' spring forwards and pulling it backwards sends a longitudinal wave through the spring

● **Sound** is an example of a longitudinal wave.

Describing waves

Revised

Waves are described in terms of wavelength, frequency and amplitude.

● **Wavelength** — the distance from a point on one wave to the corresponding point on the next wave.

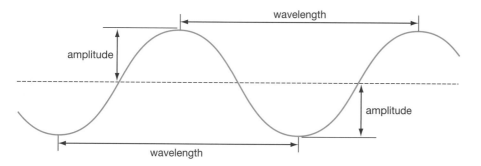

- **Amplitude** — the maximum displacement of a wave from its middle position.
- **Frequency** — the number of waves produced each second, or the number of waves that pass a point each second.

All waves obey the wave equation

Revised

$v = f \times \lambda$

wave speed = frequency × wavelength

- v in metres/second, m/s
- f in hertz, Hz
- λ in metres, m

examiner tip

You don't need to remember the wave equation — but in a higher-tier question you may need to be able to rearrange it and remember the units.

Sound

Revised

- When an object vibrates, it makes the air next to it vibrate. When the vibrations reach your ears, your ear drum vibrates and you hear sound.
- Sound waves cannot travel through a **vacuum** — a vacuum is empty space — without particles, the energy from a vibrating object cannot be transferred.
- Musicians often describe notes in terms of **pitch**. The pitch of a note depends on the frequency of the sound waves — low-frequency notes have a lower pitch than high-frequency notes.
- An **echo** is a reflected sound wave — the reflected sound is heard a short time after the original sound.

Check your understanding

Tested

40 Radio waves, used to communicate with a submarine are transmitted through the air at a frequency of 3 kHz. Calculate the wavelength of the waves as they travel through the air at 3×10^8 m/s. *(2 marks)*

41 Give two examples of transverse waves. *(2 marks)*

42 Explain how a transverse wave is different from a longitudinal wave. *(2 marks)*

43 Water waves are produced with a wavelength of 2 cm. The waves travel at 0.4 m/s. Calculate the frequency of the oscillation producing the waves. *(2 marks)*

44 Why are you more likely to hear an echo standing near a tall building than you are standing in an open field? *(1 mark)*

Answers online **Test yourself online** Online

Reflection, refraction and diffraction

Reflection, **refraction** and **diffraction** can be shown using water waves in a ripple tank.

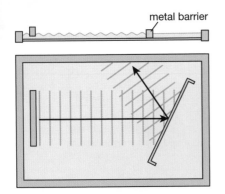

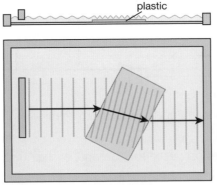

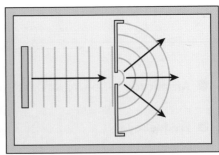

↑ The water waves reflect off the barrier at the same angle as they hit it

↑ As they enter the shallower water above the plastic sheet, the water waves change direction — the waves have been refracted

↑ When the water waves pass through the gap in the barrier they spread out — the waves have been diffracted

Reflecting light

Revised ☐

- **Reflection** is the change in direction of a wave when it hits a surface.

A mirror with a flat surface is called a plane mirror.

- **normal** — a line drawn at 90° to a surface
- **angle of incidence** (i) — the angle between the incident ray and normal
- **angle of reflection** (r) — the angle between the reflected ray and normal

When light is reflected at a surface, the light rays obey the rule:

angle of incidence (i) = angle of reflection (r)

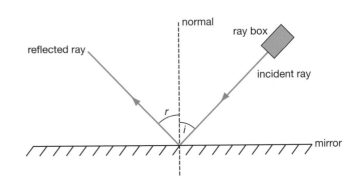

↑ Reflection in a plane mirror

Image in a plane mirror

Revised ☐

The light reflected from the mirror seems to come from an image behind the mirror — this is a **virtual image**.

The image in a plane mirror is always:

- virtual
- upright
- laterally inverted (turned sideways)

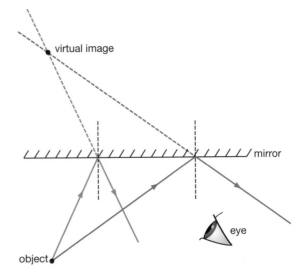

examiner tip

If you are asked to draw a ray diagram, make sure you use a ruler.

↑ When you look in a plane mirror, the image you see appears to be behind the mirror

Refraction of light

Revised

● **Refraction** — the change in direction of a wave as it passes across the boundary from one medium to another.

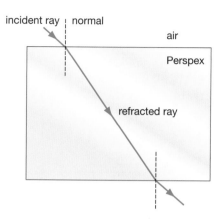

incident ray | normal

air

Perspex

refracted ray

← The light is refracted when it enters and when it leaves the block — the ray of light entering the block and the ray leaving are parallel

Diffraction

Revised

● **Diffraction** — the spreading out of a wave as it passes through the gap in a barrier or moves past an obstacle.

Diffraction is most obvious when the width of the gap is about the same as the wavelength of the wave.

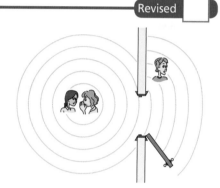

→ **The sound waves diffract and spread out as they pass through the open doorway. The light waves do not diffract at the doorway — it is millions of times wider than the wavelength of light**

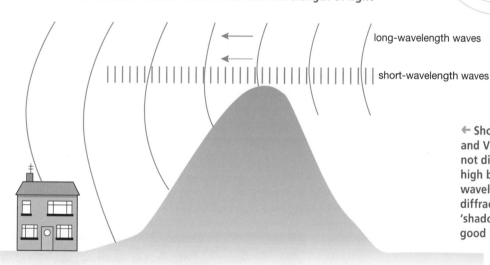

long-wavelength waves

short-wavelength waves

← Short-wavelength television and VHF radio waves are not diffracted by hills or high buildings, but longer-wavelength radio waves are diffracted. So even in the 'shadow' of a hill there can be good long-wave radio reception

Check your understanding

Tested

45 An object is 3 cm in front of a plane mirror. Draw a ray diagram to show how the image is formed behind the mirror. *(3 marks)*

46 Which way does light refract when it goes from air into water — away from or towards the normal? *(1 mark)*

47 Suggest a suitable width for the gap in a barrier in order to show the diffraction of light waves. Give a reason for your answer. *(2 marks)*

48 How can you deduce that both transverse and longitudinal waves can be diffracted? *(1 mark)*

Answers online Test yourself online Online

Electromagnetic waves and communications

Electromagnetic radiations travel as waves
Revised

Electromagnetic waves are **transverse** and, like all waves, transfer energy from one place to another.

Electromagnetic waves cover a continuous range of wavelengths forming a continuous spectrum — the **electromagnetic spectrum**.

The **range** of wavelengths goes from approximately 10^{-15} m to more than 10^4 m.

Seven types of electromagnetic wave make up the electromagnetic spectrum.

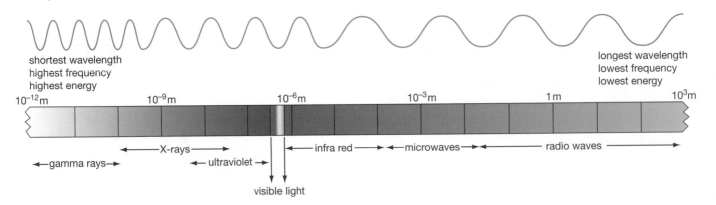

shortest wavelength
highest frequency
highest energy

longest wavelength
lowest frequency
lowest energy

10^{-12} m 10^{-9} m 10^{-6} m 10^{-3} m   1 m 10^3 m

←gamma rays→   ←X-rays→   ← ultraviolet →   ← infra red →   ←microwaves→   ← radio waves →

visible light

Electromagnetic waves have some common properties:

● They all travel at the same speed through a vacuum.

● They obey the wave equation.

● They can be reflected, refracted and diffracted.

> **examiner tip**
>
> Learn the order of the waves going from lowest energy (radio) to highest energy (gamma).

Communications
Revised

Visible light, infra red, microwaves and radio waves can be used for communications. Visible light and infra red signals can be sent along an **optical fibre**. The signal travels from one end of the fibre to the other by repeated reflections.

Visible light is used to produce images in photography. Remote television controls use infra red to carry signals over short distances.

Microwaves are used in:

● **satellite** communication systems — microwaves pass easily through the Earth's atmosphere

● **mobile phone** networks — signals are transmitted over long distances via tall aerial masts or long distances via a satellite

Radio waves are transmitted over long distances by reflecting them off the ionosphere (a layer of ionised gas in the Earth's upper atmosphere).

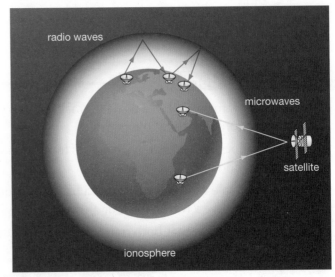

radio waves

microwaves

satellite

ionosphere

Hazards from infra red and microwaves

- Infra red — when absorbed by soft body tissue it is felt as heat; infra red can cause burns.

- Microwaves — living cells contain water so are heated by microwaves; this may damage or kill the cells

Mobile phone networks use microwaves. We know that, when absorbed, microwaves will heat body cells because the cells contain water. However, at present the long-term effects of exposure to microwaves due to using a mobile phone or living near a phone mast are unknown.

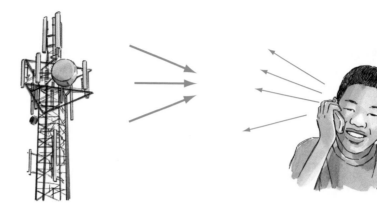

Recent research involving more than 4000 people concluded that, in the first 10 years of using a mobile phone, there was no increased risk of cancer. But long-term risks remain **unknown**.

Check your understanding

49 Which part of the electromagnetic spectrum is used:

  a)   in mobile phone networks?

  b)   to communicate with a satellite?

  c)   to operate a television remote control?   *(3 marks)*

50 Which parts of the electromagnetic spectrum have a lower frequency than visible light?   *(1 mark)*

51 State two ways that infra red waves are different from radio waves.   *(2 marks)*

52 Why is it important that scientists continue to monitor the health of mobile phone users?   *(2 marks)*

53 The table gives the specific absorption rate (SAR) value for three different mobile phones. The SAR value measures the radiation energy absorbed by the head when a mobile phone is used.

Mobile phone	SAR value
X	1.41
Y	0.69
Z	0.22

SAR values are measured in the laboratory to give the maximum rate of energy absorption. The SAR value can be much lower when the phone is actually used.

  a)   Why is it better to give the SAR value obtained in the laboratory rather than a value obtained when the phone is in use?   *(2 marks)*

  b)   The maximum SAR value that a phone sold in Europe can have is 2.0. Does this mean the three phones (X, Y and Z) are safe to use? Give the reason for your answer.   *(1 mark)*

  c)   Some parents are going to buy a mobile phone for one of their children. Which phone, X, Y or Z, would you recommend that they buy? Explain the reason for your choice.   *(2 marks)*

Answers online   Test yourself online

Expanding universe and 'big bang'

The Doppler effect

The Doppler effect happens with all waves, but is most easily noticed with **sound**. If a source of sound of constant frequency moves away from you, the frequency of the sound you hear decreases. This happens because the waves seem to stretch, making the observed wavelength of the waves longer.

In a similar way, if the source of the sound moves towards you, the frequency you hear increases and the waves appear to have a shorter wavelength.

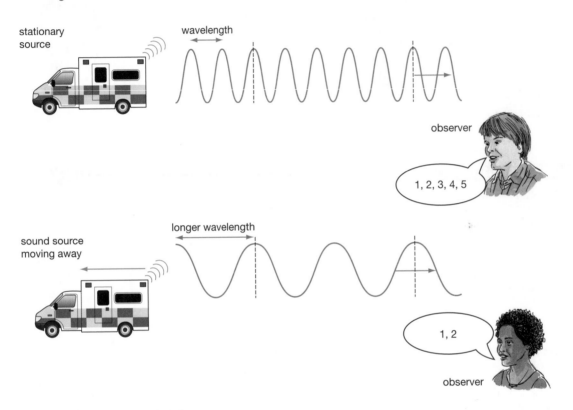

What is red-shift?

If a source of light moved away from you quickly enough, the light waves you see would have a longer wavelength and decreased frequency. This would make the light appear as if it has moved towards the red end of the spectrum. This effect is called red-shift.

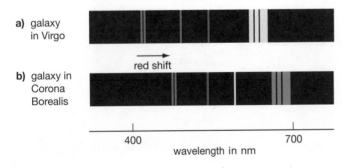

a) galaxy in Virgo

red shift

b) galaxy in Corona Borealis

400   700

wavelength in nm

examiner tip

Red-shift means that wavelengths appear longer — it does not mean that a galaxy looks red or emits only red light.

Expanding universe

Revised

The 'big bang' theory suggests that the universe began with a huge explosion, billions of years ago. At that moment, all the matter in the universe, concentrated at a tiny point in space, started to expand outwards into the universe we have today. This is only a theory — scientists cannot prove the theory, but they can look for evidence to support it.

Evidence for a rapidly expanding universe

Revised

When scientists look at the light from distant **galaxies**, it shows red-shift. So the galaxies must be **moving away** from the Earth at great speed. Scientists have observed that the further a galaxy is from the Earth, the bigger the red-shift. This observation can be explained by suggesting that galaxies further away must be moving more quickly.

So, at some time in the past they must have been closer together. Red-shift provides evidence to support the 'big bang' theory.

The 'big bang' theory is not the only way of explaining how the universe began. Currently it is the most popular — but ideas can change. New evidence may mean adapting or even rejecting the 'big bang' theory. A theory is only as good as the evidence that supports it.

Cosmic microwave background radiation

Revised

Cosmic microwave background radiation (CMBR), provides further evidence for the 'big bang' theory.

CMBR — created just after the 'big bang' — started as short wavelength gamma radiation. As the universe has expanded, the wavelength has increased so that it is now detected as microwave radiation.

Currently only the 'big bang' theory can explain the existence of CMBR.

Check your understanding

Tested

54 Explain why red-shift provides evidence to support the 'big bang' theory. *(3 marks)*

55 Which one of the following statements about the 'big bang' theory is correct? *(1 mark)*

    A It is the only way of explaining the origin of the universe.

    B There is scientific evidence to support the theory.

    C It is based on scientific and religious facts.

    D Scientists have proof that it happened.

56 Name one piece of evidence, other than red-shift, that supports the 'big bang' theory. *(1 mark)*

Answers online — Test yourself online — Online

Forces and resultant forces

Forces always come in pairs

When two bodies act on each other, the forces they exert on each other are always equal in size and opposite in direction.

Because each force acts on a different body, they do not cancel out.

Two or more forces acting on a body give a resultant force

Imagine that all the forces on a body replaced are by a single force that has the same effect — this single force is the **resultant force**.

push up from the skateboard

push down on the skateboard

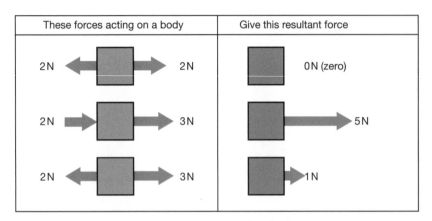

These forces acting on a body	Give this resultant force
2N ← □ → 2N	□ 0N (zero)
2N → □ → 3N	□ → 5N
2N ← □ → 3N	□ ► 1N

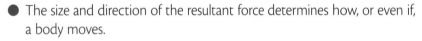

● The size and direction of the resultant force determines how, or even if, a body moves.

A zero resultant force does not change movement

● If a body is **stationary** (not moving), it will remain stationary.

● If a body is moving, it will carry on moving at the same speed and in the same direction (**constant velocity**).

A non-zero resultant force always changes movement

● If the resultant force on a body is not zero, the body will accelerate in the direction of the resultant force.

   a) The stationary car starts to accelerate in the direction of the resultant force.

   b) The resultant force is in the same direction as the car is moving — so the car accelerates.

   c) The car is moving in the opposite direction to the resultant force — so the car decelerates (slows down).

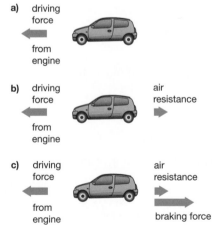

a) driving force from engine

b) driving force from engine   air resistance

c) driving force from engine   air resistance   braking force

A force can change the shape of an object. If a force is applied to an **elastic object** the shape of the object changes, but not permanently. Removing the force lets the object go back to its original shape.

The **force** exerted on a spring and the **extension** (how much stretch) of the spring are linked by the equation:

$F = k \times e$

force = spring constant × extension

● F in newtons, N

● k in newtons per metre, N/m

● e in metres, m

This equation works provided the spring has not gone past its **limit of proportionality** — it has not been overstretched.

When an elastic object is stretched, squashed, twisted or bent it stores **elastic potential energy** — this is the energy stored in an elastic object when work is done to change its shape.

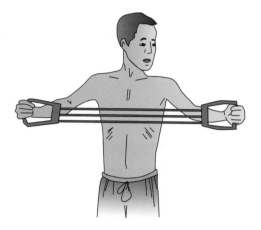

↑ **When released, the springs go back to their original size — they were not stretched beyond their limit of proportionality**

Check your understanding

Tested ▢

1 The diagram shows the forces acting on a flying aircraft. Describe the motion of the aircraft when:

   a) lift = weight   (1 mark)

   b) thrust = drag   (1 mark)

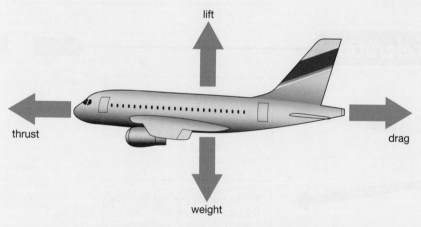

2 What is the resultant force on a motorcyclist travelling at constant velocity?   (1 mark)

3 A force of 1200 N stretches a strong spring by 5 cm. Calculate, in N/m, the spring constant for the spring.   (2 marks)

4 How would you be able to tell that a force has *not* stretched a spring beyond its limit of proportionality?   (1 mark)

Answers online ──── **Test yourself online** ──────────── Online ▢

Speed, velocity and acceleration

Speed and velocity Revised

The **speed** of a car, measured in metres per second (m/s), is how far it travels in 1 second.

Velocity is also measured in metres per second. But velocity is not the same as speed — when quoting a velocity you must give both speed and direction.

● The velocity of a body is its speed in a given direction.

↑ The two cars have the same speed, but in opposite directions — so they have different velocities

Velocity and acceleration Revised

A body is accelerating when its velocity is changing. The faster the velocity changes, the larger the **acceleration**.

Acceleration is calculated using the equation:

$$a = \frac{v - u}{t}$$

$$\text{acceleration} = \frac{\text{change in velocity}}{\text{time taken for change}}$$

● a in metres per second squared, m/s^2

● v and u in metres per second, m/s

● t in seconds, s

If your answer for acceleration is negative, the body is slowing down — it is **decelerating**.

> **examiner tip**
> The units for velocity and acceleration look similar and are easily confused. Remember acceleration includes the number 2: m/s^2.

Resultant force and acceleration Revised

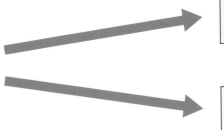

A body will accelerate when the resultant force acting is not zero.

The greater the resultant force on the body, the greater the acceleration of the body.

The bigger the mass of the body, the bigger the resultant force needed to make it accelerate.

The acceleration of an object is calculated using the equation:

$$a = \frac{F}{m}$$

$$\text{acceleration} = \frac{\text{resultant force}}{\text{mass}}$$

● F in newtons, N

● a in metres per second squared, m/s^2

● m in kilograms, kg

Distance–time graphs

Revised

The slope of a **distance–time graph** represents speed.

- A horizontal line means the body is not moving.
- A straight line going up means the body is moving at constant speed.
- The steeper the slope of the line, the faster the speed.

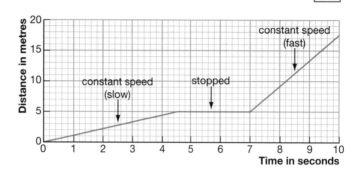

Velocity–time graphs

Revised

The slope of a **velocity–time graph** represents acceleration.

The area under a velocity–time graph represents the distance travelled.

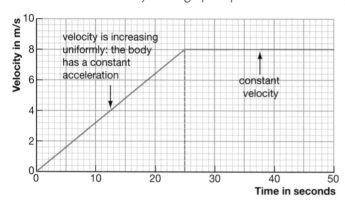

> **examiner tip**
>
> Always check the label on the *y*-axis to see if it is a distance–time graph or a velocity–time graph. The graphs may look the same but they mean different things.

- A horizontal line means the body is not accelerating — it is moving at constant velocity.
- The steeper the slope of the line, the greater the acceleration.
- The bigger the area under the line, the greater the distance travelled.

Check your understanding

Tested

5 The velocity of a speed skier down a straight, steep slope increases from 12 m/s to 32 m/s in 4 seconds. Calculate the acceleration of the skier. *(2 marks)*

6 Draw the distance–time graph for a jogger running at constant speed along a straight road. *(2 marks)*

7 A horse running on a straight track accelerates uniformly from 0 m/s to 12 m/s in 6 seconds. It gallops at this speed for 15 seconds, before slowing down uniformly, and coming to a halt in a further 4 seconds.

   a) Calculate the acceleration of the horse in the first 6 seconds. Show your working. *(2 marks)*

   b) Calculate the deceleration of the horse in the last 4 seconds. *(2 marks)*

   c) Draw a velocity–time graph for the horse. *(4 marks)*

8 The speed of an 800 kg car increases from 5 m/s to 15 m/s in 20 s.

   a) Calculate the acceleration of the car. *(2 marks)*

   b) Calculate the resultant force needed to produce the acceleration. *(2 marks)*

Answers online — **Test yourself online** Online

More on speed, velocity and acceleration

Calculating the gradient (slope) Revised

Remember — you must use the numbers on the graph scales to work out the values of x and y.

Example: what is the gradient of this graph line?

$y = 8 - 2 = 6$

$x = 4 - 1 = 3$

$\text{slope} = \dfrac{y}{x} = \dfrac{6}{3} = 2$

→ **Calculating the gradient of a straight-line graph**

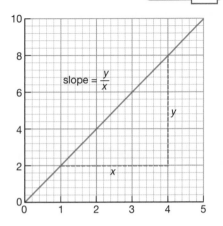

slope $= \dfrac{y}{x}$

Distance–time graphs: calculating speed Revised

Calculating the gradient of the distance–time graph for an object gives the speed of the object.

Example: The distance–time graph below shows a small part of a cyclist's journey. Calculate the slowest speed of the cyclist.

$y = 600 - 400 = 200$

$x = 100 - 50 = 50$

$\text{slope} = \dfrac{y}{x} = \dfrac{200}{50} = 4\,\text{m/s}$

You should be able to show that the fastest speed of the cyclist was 8 m/s.

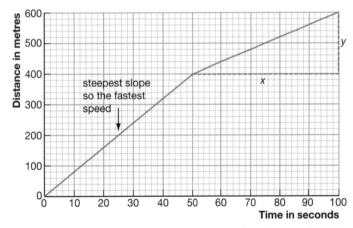

steepest slope so the fastest speed

examiner tip

Always remember to give your answer a unit — it may be worth a mark.

Velocity–time graphs: calculating acceleration Revised

● Calculating the gradient of the velocity–time graph for an object gives the acceleration of the object.

● Calculating the area under the line on a velocity–time graph gives the distance travelled by the object.

Example: The velocity–time graph on page 37 is for a motorcyclist travelling along a straight road.

a) Calculate the initial acceleration of the motorcyclist.

$\text{acceleration} = \text{slope of the first line} = \dfrac{12}{8} = 1.5\,\text{m/s}^2$

You should be able to show that the motorcyclist finally decelerates at 1.2 m/s².

b) Calculate the total distance travelled by the motorcyclist.

Divide the graph into three parts, calculate the area of each part and add together:

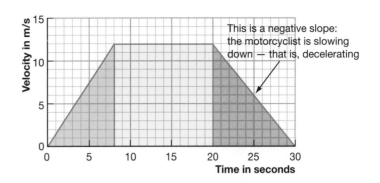

$$= (\frac{1}{2} \times 8 \times 12) + (12 \times 12) + (\frac{1}{2} \times 10 \times 12)$$

$$= 48 + 144 + 60$$

$$= 252 \, m$$

Check your understanding
Tested

9 The graph shows how the distance walked by a person changes with time.

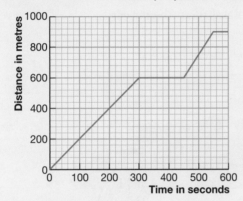

   a) Calculate the fastest speed that the person walked at. *(3 marks)*

   b) Calculate the slowest speed that the person walked at. *(3 marks)*

10 The graph shows how the velocity of a small aircraft changes as it accelerates down the runway and takes off.

   Calculate:

   a) the acceleration of the aircraft *(2 marks)*

   b) the length of runway used for take-off *(3 marks)*

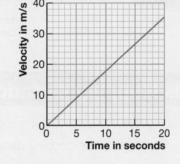

11 The graph shows how the velocity of a lift changes between two floors.

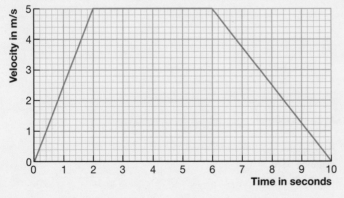

   Calculate:

   a) the deceleration of the lift *(2 marks)*

   b) the distance between the two floors *(3 marks)*

Answers online — Test yourself online — Online

Forces and braking

When a vehicle travels at a constant speed, the resistive forces — mainly air resistance — balance the driving force produced by the engine.

The work done by a braking force reduces a vehicle's kinetic energy. As the kinetic energy goes down, the temperature of the brakes goes up.

Stopping distance
Revised

- A total **stopping distance** has two parts:

  stopping distance = thinking distance + braking distance

- The **thinking distance** is how far a vehicle travels in the time it takes the drivver to react to a hazard and apply the brakes — i.e. the driver's reaction time.

- The **braking distance** is how far a vehicle travels before stopping, once the brakes have been applied.

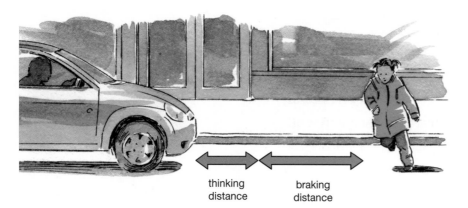

thinking distance   braking distance

Speed changes stopping distance
Revised

If the force applied by the brakes is the same at every speed, then:

- The faster the vehicle, the longer is the stopping distance.

At 13 m/s (30 mph)

Thinking   **Braking**   Overall stopping
distance 9 m   distance 14 m   distance 23 m

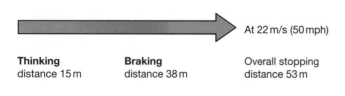

At 22 m/s (50 mph)

Thinking   **Braking**   Overall stopping
distance 15 m   distance 38 m   distance 53 m

At 31 m/s (70 mph)

Thinking   **Braking**   Overall stopping
distance 21 m   distance 75 m   distance 96 m

↑ **Stopping distances for a typical car in good condition on a dry road**

Stopping quickly means rapid deceleration

A large braking force is needed for a vehicle to decelerate rapidly and stop over a short distance.

● The faster the vehicle, the greater the braking force needed to make the vehicle stop in the same distance.

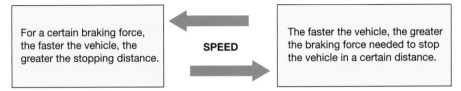

| For a certain braking force, the faster the vehicle, the greater the stopping distance. | ← **SPEED** → | The faster the vehicle, the greater the braking force needed to stop the vehicle in a certain distance. |

Speed is important but it is not the only factor to affect stopping distance.

Thinking distance is affected by:

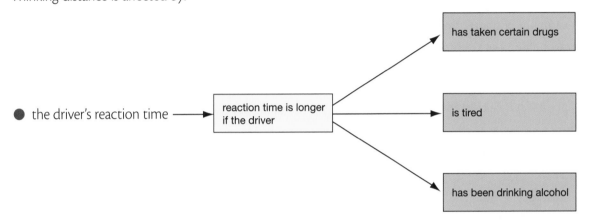

● the driver's reaction time → reaction time is longer if the driver → has taken certain drugs / is tired / has been drinking alcohol

A driver's ability to react may also be affected by distractions, such as talking on a mobile phone.

Braking distance is affected by speed and by:

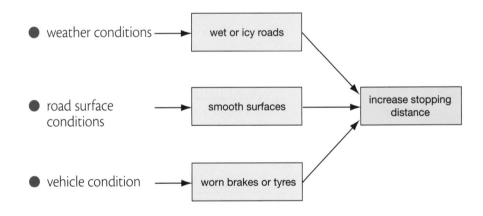

● weather conditions → wet or icy roads

● road surface conditions → smooth surfaces → increase stopping distance

● vehicle condition → worn brakes or tyres

examiner tip

Factors that affect reaction time are to do with the driver. Factors that affect braking distance are to do with the weather, the road or the car.

Check your understanding

12 Why is the thinking distance for stopping a car unaffected by there being ice on the road? *(1 mark)*

13 A motorcyclist is travelling at 18 m/s. Write down three factors that could affect the stopping distance of the motorcyclist. *(3 marks)*

14 Suggest why having noisy car passengers may result in an increased stopping distance. *(2 marks)*

Answers online **Test yourself online** Online ☐

Forces and terminal velocity

Frictional forces

Revised ☐

- Frictional forces usually act on moving bodies.
- Frictional forces always act on an object that is moving through a **fluid**.
- Frictional forces oppose the movement of the object.
- A **fluid** is a substance that flows — liquids and gases are fluids.
- The faster an object moves through a fluid, the bigger the frictional force becomes.
- **Weight** is the force exerted by gravity on an object — it is not the same as the mass of the object.
- **Mass** is the amount of matter that makes up an object.

On Earth, gravity pulls on every 1 kilogram of mass with a force of about 10 newtons. This is called the **gravitational field strength** (g). On Earth, $g = 10\,\text{N/kg}$.

The weight of an object is calculated using the equation:

$$W = m \times g$$

weight = mass × gravitational field strength

- W in newtons, N
- m in kilograms, kg
- g in newtons per kilogram, N/kg

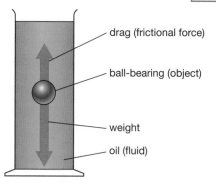

drag (frictional force)

ball-bearing (object)

weight

oil (fluid)

↑ **The forces on a ball-bearing falling through a fluid**

> **examiner tip**
> Different names such as air resistance, drag and water resistance are used to describe frictional forces; but they all do the same thing — try to stop a body from moving.

> **examiner tip**
> An arrow labelled 'weight' is usually used in diagrams, not an arrow labelled 'force of gravity'.

Frictional forces change with velocity

Revised ☐

backward frictional force (air resistance)

forward force from cyclist

← **The forward and backward forces acting on a cyclist**

> **examiner tip**
> It doesn't matter what the object is, if it has reached its terminal velocity then the resultant force must be zero.

The graph shows how the velocity of a cyclist changes with time.

A When the cyclist starts moving forwards, a frictional force acts on the cyclist backwards.

The force forwards is larger than the frictional force (air resistance) backwards, so the cyclist's velocity increases.

B As velocity increases, air resistance increases.

The resultant force decreases, reducing the acceleration of the cyclist. (So, the gradient of the graph decreases.)

C The air resistance increases until it equals the forward force.

The resultant force is zero and the cyclist stops accelerating (the gradient of the graph is zero). The cyclist continues to move but at a constant velocity called the **terminal velocity**.

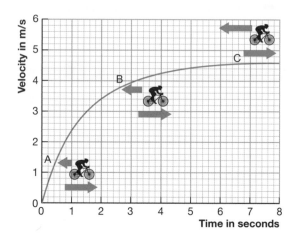

↑ **The velocity–time graph for a cyclist**

Forces on a parachutist

By increasing the drag force, a parachute reduces the terminal velocity of a person falling through the air.

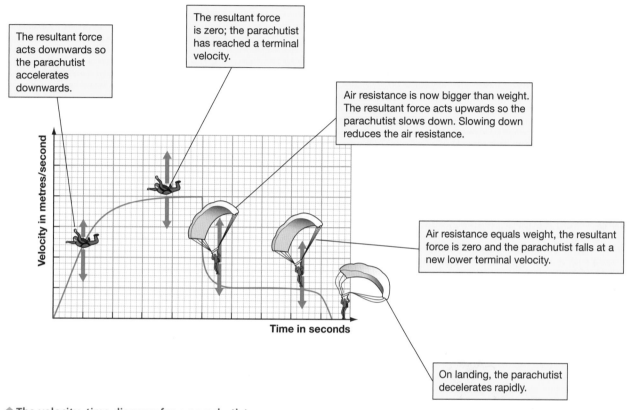

The resultant force acts downwards so the parachutist accelerates downwards.

The resultant force is zero; the parachutist has reached a terminal velocity.

Air resistance is now bigger than weight. The resultant force acts upwards so the parachutist slows down. Slowing down reduces the air resistance.

Air resistance equals weight, the resultant force is zero and the parachutist falls at a new lower terminal velocity.

On landing, the parachutist decelerates rapidly.

⬆ The velocity–time diagram for a parachutist

Check your understanding

15 A book has a mass of 430 g. How much does the book weigh? *(2 marks)*

16 An astronaut weighs 800 N on the Earth.

  a) What is the mass of the astronaut? *(2 marks)*

  b) What would the astronaut weigh on the Moon? The gravitational field strength on the Moon is 1.6 N/kg. *(2 marks)*

17 A cyclist travelling at a constant velocity produces a forward force of 50 N.

  a) What name is given to the constant velocity of the cyclist? *(1 mark)*

  b) What is the total backward force acting on the cyclist? *(1 mark)*

  c) What is the resultant force on the cyclist? *(1 mark)*

18 A golf ball and a table tennis ball are dropped from the same height. The golf ball hits the ground before the table tennis ball.

  a) Why does the golf ball hit the ground first? *(1 mark)*

  b) How would the motion of the golf ball and table tennis ball change if they were dropped in a vacuum on the Earth? *(1 mark)*

Answers online **Test yourself online**
Online

Forces and energy

Whenever a force moves an object, energy is transferred and **work** is done.

The amount of work done depends on:

● the size of the force

● the distance the force moves

Work done is calculated using the equation:

$$W = F \times d$$

work done = force applied × distance moved in direction of force

● W in joules, J

● F in newtons, N

● d in metres, m

Energy is needed to do work. Work and energy are both measured in joules (J).

Imagine pushing a heavy crate across the floor. The work done in pushing the crate equals the energy transferred.

The work you do to push the crate at a constant speed is against the frictional forces that are trying to stop the crate moving. As the crate rubs against the floor, the crate and the floor get hotter.

Work done against frictional forces causes energy to be transferred, increasing the temperature of the moving object.

pushing force

friction force

Work, energy transfer and power

Revised

It is often not how much work is done (or energy transferred) but how quickly the work is done that is important.

The rate at which work is done, or energy is transferred, is called **power**.

Power is calculated using the equation:

$$P = \frac{E}{t}$$

$$\text{power} = \frac{\text{energy transferred}}{\text{time taken}}$$

● P in watts, W

● E in joules, J

● t in seconds, s

Gravitational potential energy

Revised

To lift an object, work is done against gravity and the **gravitational potential energy** (GPE) of the object increases.

The change in GPE of the object is calculated using the equation:

$E_p = m \times g \times h$

change in GPE = mass × gravitational field strength × change in height

- E_p in joules, J
- m in kilograms, kg
- g in newtons per kilogram, N/kg
- h in metres, m

→ When the box is lifted, work is done — the box gains GPE equal to the work done by the force that lifts the box

Kinetic energy

Revised

The **kinetic energy** of a moving object depends on:

- its speed
- its mass

Kinetic energy is calculated using the equation:

$E_k = \dfrac{1}{2} \times m \times v^2$

kinetic energy = $\dfrac{1}{2}$ × mass × speed squared

- E_k in joules, J
- m in kilograms, kg
- v in metres per second, m/s

examiner tip
Remember — in this equation its only speed that is squared, not everything on the right-hand side. Also practise rearranging the equation — you could be asked to calculate mass or speed.

Regenerative braking

Revised

Braking and then accelerating a vehicle means that energy stored in the fuel is used to replace the energy lost by braking — this is inefficient.

With an electric or hybrid car, some of the kinetic energy lost when the brakes are applied is used to make a generator work to charge the car's battery. This system, called **regenerative braking**, increases efficiency because some of the energy that would have been lost has been stored for later use.

Check your understanding

Tested

19 A shopper does 1000 J of work pushing a loaded trolley 20 m at a constant speed.
   a) Calculate the force used to push the trolley. *(2 marks)*
   b) What happens to the energy transferred from the shopper? *(1 mark)*
20 A rhinoceros moving at 4 m/s has 12 000 J of kinetic energy. Calculate the mass of the rhinoceros. *(2 marks)*
21 Calculate the kinetic energy of:
   a) a racing car of mass 500 kg travelling at 85 m/s *(2 marks)*
   b) a 40 g bird flying at 30 m/s *(2 marks)*
   c) a 25 kg dog running at 6 m/s *(2 marks)*
22 A child throws a ball of mass 0.2 kg, 3 metres up into the air. Calculate the gravitational potential energy gained by the ball. (g = 10 N/kg) *(2 marks)*

Answers online — Test yourself online — Online

Momentum

All moving bodies have momentum

The greater the **mass** of the body and the faster its **velocity**, the more **momentum** it has.

Momentum is calculated using the equation:

$p = m \times v$

momentum = mass × velocity

- p in kilogram metres per second, kg m/s
- m in kilograms, kg
- v in metres per second, m/s

mass = 18 000 kg
velocity = 8 m/s

mass = 1200 kg
velocity = 30 m/s

↑ **Which has more momentum, the lorry or the car?**

> **examiner tip**
> Remember the unit of momentum as the unit of mass (kg) multiplied by the unit of velocity (m/s).

Like velocity, momentum has a size and direction.

If two bodies move in opposite directions, one has a positive momentum and the other a negative momentum.

When a resultant force acts on a body, the body accelerates. So, both the velocity and the momentum of the body change.

Conservation of momentum

In any collision or explosion momentum is conserved. The total momentum before a collision or explosion is equal to the total momentum after the collision or explosion.

Provided it is a **closed system** — i.e. no external forces act on the colliding or exploding bodies:

- If the total momentum of two bodies before a collision or explosion is zero, then it must be zero after the collision or explosion.
- For the momentum of two moving bodies to be zero, they must be moving in opposite directions.

Car safety and momentum

A small force acting for a long time can cause the same change in momentum as a large force acting for a small time.

This principle is behind many of the features of a car that are designed to reduce injury during a collision.

- In a collision, the **crumple zones** of a car crush.
- This makes the car slow down and stop more gradually.
- The longer it takes to reduce the momentum, the smaller the force exerted.
- The smaller the force, the less likely the car occupants are to suffer serious injury.

You can also explain the way safety features work in terms of energy. For example, in a collision a seat belt is designed to stretch.

> **examiner tip**
>
> Make sure you can use the idea of 'the longer the time to change momentum means a smaller force exerted' to explain other safety features, such as air bags, seatbelts and side impact bars.

| **Kinetic energy lost** by the person wearing the seat belt | **=** | **Work done** in stretching the seat belt |

Check your understanding

23 Copy and complete the table. *(3 marks)*

	Mass	**Velocity**	**Momentum**
Horse	600 kg	8 m/s	
Jogger		6 m/s	450 kg m/s
Toy car	60 g		0.15 kg m/s

24 A 0.3 kg trolley is moving at 2 m/s and it collides with and sticks to a stationary 0.1 kg trolley. Calculate:

   a) the total momentum before the collision   *(2 marks)*

   b) the velocity of the two trolleys after the collision.   *(3 marks)*

25 Explain how an air bag helps in reducing the risk of injury to a car driver during a collision.   *(3 marks)*

26 A 600 kg, high-performance car travelling at 30 m/s is brought to a halt in 4 s. Calculate the braking force applied to the car.   *(3 marks)*

Answers online   **Test yourself online**   Online

Static electricity

Charging by friction

Revised

- Certain insulating materials become **electrically charged** when rubbed together.
- Friction causes negative electrons to move from one material to the other.
- The material **gaining electrons** becomes **negatively** charged.
- The material **losing electrons** is left with a **positive** charge.

Charging a polythene rod

The polythene rod gains electrons from the cloth. The rod becomes negatively charged.

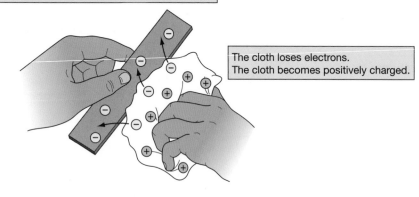

The cloth loses electrons. The cloth becomes positively charged.

Charging a Perspex rod

The Perspex rod loses electrons to the cloth. The rod becomes positively charged.

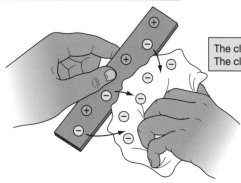

The cloth gains electrons. The cloth becomes negatively charged.

> **examiner tip**
>
> In any question about charging by friction, it is always the negatively charged electrons that do the moving.

Charged objects exert forces on each other

Revised

- Objects with the **same charge repel** (+ and +, or − and −).
- Objects with **opposite charge attract** (+ and −).

If the objects are close enough, the **electrostatic force** may make them move.

→ Because the balloons have the same type of charge, they repel and move away from each other

Charge moves easily through a conductor

An electric **current** is a flow of **charge**.

● When the conductor is a solid, the current is a flow of **electrons**.

● The size of an electric current is the rate of flow of charge.

The greater the amount of charge flowing through a circuit each second, the bigger the current. The size of an electric current is calculated using the equation:

$$I = \frac{Q}{t}$$

$$\text{current} = \frac{\text{charge}}{\text{time}}$$

● Q in coulombs, C

● I in amps, A

● t in seconds, s

1 coulomb is the amount of charge that passes a point in a circuit when a current of 1 ampere flows for 1 second.

Potential difference and charge

Charge only flows between two points in a circuit if there is a **potential difference** between the two points.

The **potential difference** is equal to the work done (or energy transferred) for each coulomb of charge passing between the two points.

Potential difference is calculated using the equation:

$$V = \frac{Q}{t}$$

$$\text{potential difference} = \frac{\text{charge}}{\text{time}}$$

● Q in coulombs, C

● V in volts, V

● t in seconds, s

> **examiner tip**
>
> Potential difference is also called voltage. In exam questions the term potential difference will be used.

Check your understanding

27 Will two charged ebonite rods attract or repel? Give a reason for your answer. *(2 marks)*

28 Pete rubs his jumper with a balloon. The balloon becomes negatively charged.

   a) Explain why the balloon becomes negatively charged. *(2 marks)*

   b) What happens to the jumper? *(2 marks)*

29 What is an electric current? *(1 mark)*

30 An electric shower draws a current of 35 A from the mains supply. A total charge of 10 500 C passes through the shower circuit while it is switched on. Calculate how long the shower is switched on for. *(2 marks)*

Answers online   **Test yourself online**

Electrical circuits

Standard symbols for electrical components

A circuit diagram uses the standard symbols to show how components are joined together.

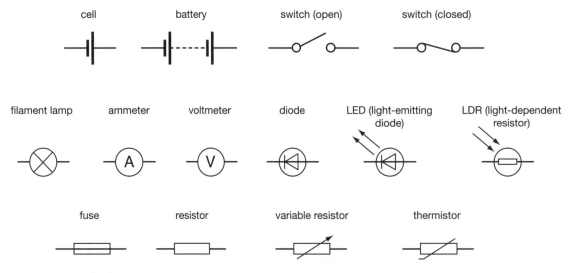

cell   battery   switch (open)   switch (closed)

filament lamp   ammeter   voltmeter   diode   LED (light-emitting diode)   LDR (light-dependent resistor)

fuse   resistor   variable resistor   thermistor

> **examiner tip**
> Learn these circuit symbols — you may be asked to draw and/or interpret circuit diagrams using them.

Batteries

A battery is two or more cells joined in series. The total potential difference of cells joined in **series** is worked out by adding the separate potential differences together. This only works if the cells are joined positive (+) to negative (−).

All components have a resistance

The easier a current flows through a component, the less **resistance** the component has.

- Resistance is measured in **ohms** (Ω).
- An **ammeter** measures the **current** (I) through the component.
- A **voltmeter** measures the **potential difference** (V) across the component.

Provided the potential difference across the component does not change, increasing the resistance of a component will decrease the current flowing through the component.

Potential difference, current and resistance are linked by the equation:

$$V = I \times R$$

potential difference = current × resistance

- V in volts, V
- I in amps, A
- R in ohms, Ω

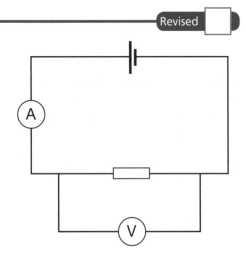

Series circuits

Components in a series circuit are joined in a continuous loop. The current in a **series circuit** has only one path it can take. This means that:

● the same current (I) flows through each component

● the potential difference of the power supply is shared between the components

● adding more components increases the resistance of the circuit.

The total resistance is worked out by adding the resistances of all the individual components.

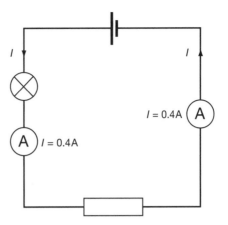

Parallel circuits

Current in a parallel circuit has more than one pathway. For components joined **in parallel**:

● the potential difference across each component is the same

● the bigger the resistance of a component, the smaller the current flowing through it

● the total current flowing in the whole circuit is worked out by adding the currents through the individual parallel components.

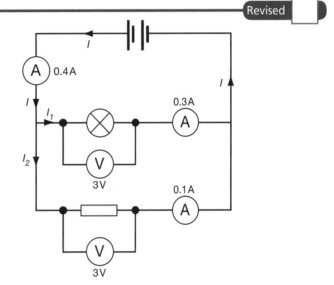

Check your understanding

31 Work out the total resistance of these arrangements.   (*3 marks*)

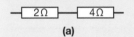

(a)

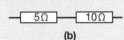

(b)

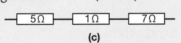

(c)

32 Copy and complete the table.   (*6 marks*)

Appliance	Potential difference across appliance in volts	Current through appliance in amps	Resistance of appliance in ohms
Lamp	12		4
Hair dryer		5	46
Electric blanket	230	0.5	

33 Draw the circuit that a student should use to measure the resistance of a lamp.   (*3 marks*)

Resistance, current and potential difference

Current–potential difference graphs

Revised

Current–potential difference graphs are used to show how the current through a component depends on the potential difference across the component.

When the component is a resistor

Provided the temperature of the resistor is constant, the graph is always a straight line going through the origin (0, 0). This shows that as the potential difference increases, the current increases in the same proportion.

- The current through a resistor is directly proportional to the potential difference across the resistor.
- A straight-line graph shows that resistance is constant.

A metal wire, at constant temperature, has a constant resistance.

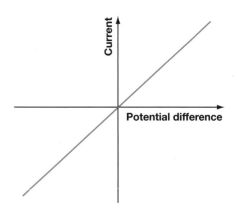

When the component is a filament lamp

The metal filament inside the lamp contains free electrons and ions. Increasing the potential difference causes the free electrons to collide more frequently with the ions and the wire gets hot. Effectively, the free electrons move more slowly around the circuit. So the resistance has increased.

The increased resistance means that the current does not increase in the same proportion as the increase in potential difference.

- The resistance of a filament lamp is not constant — it depends on the potential difference.

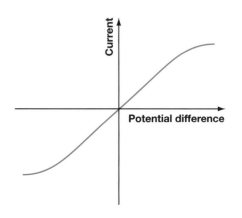

When the component is a diode

The resistance of a **diode** depends on which way round it is connected in a circuit.

- In the reverse direction, diodes have a very high resistance so no current flows.
- Diodes let current flow in only one direction.

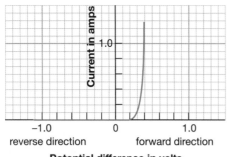

Light-emitting diodes (LEDs)

When an electric current flows through an LED in the forward direction, it emits light. Lamps that use LEDs are expensive. However, they use very little energy, have a long lifetime (about 100 000 hours) and are very reliable.

examiner tip

You should be able to identify and/or draw each of these current–potential difference graphs.

LDRs and thermistors

LDRs and thermistors are special types of resistors. The resistances of a **light-dependent resistor** (**LDR**) and a **thermistor** change in response to changes to the environment.

The resistance of an LDR goes down as the intensity of the light hitting it goes up.

● An LDR can be used in a circuit as a **light sensor**.

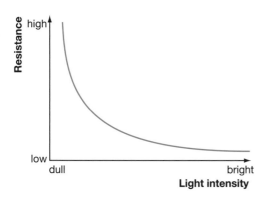

The resistance of a **thermistor** goes down as the temperature of the thermistor goes up.

● A thermistor can be used in a circuit as a **temperature sensor**.

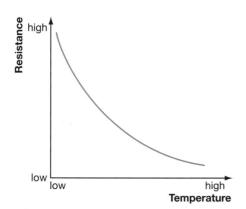

Check your understanding

34 Draw the circuit symbols for: **a)** a resistor; **b)** a lamp; **c)** a diode; **d)** an LDR; **e)** a thermistor. *(5 marks)*

35 Why is the current–potential difference graph for a filament lamp not a straight line? *(2 marks)*

36 How does the current through an LDR change with increasing light intensity? *(1 mark)*

37 The current through a thermistor at 40°C is 0.002 A. The resistance of the thermistor is 600 Ω.

   a) Calculate the potential difference across the thermistor. *(2 marks)*

   b) What happens to the resistance of the thermistor as its temperature increases? *(1 mark)*

Answers online   Test yourself online   Online

Oscilloscope traces

An **oscilloscope** (**CRO**) trace is a graph of potential difference against time.

Direct current Revised

Direct current always flows in the same direction. Cells and batteries give a **direct current (d.c.)**. The oscilloscope trace for a steady d.c. supply is a horizontal line.

The bigger the potential difference of the supply, the further the line is from the centre of the screen.

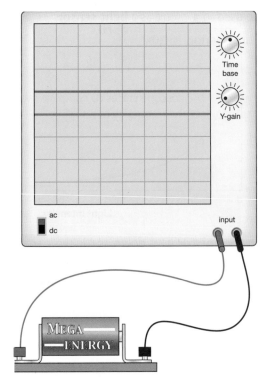

→ **Oscilloscope trace of a d.c. supply —
the zero value is shown as a red line**

Alternating current Revised

Alternating current repeatedly changes its direction. The oscilloscope trace for an **alternating (a.c.)** supply alternates between positive and negative potential difference.

The distance from the centre to the top of the trace represents the **peak** (maximum) potential difference (V_0):

● trace A is 1½ times higher than trace B
● if the peak potential difference of supply
  B = 4 V
● then the peak potential difference of supply
  A = 1½ × 4 = 6 V.

Each time the trace changes from above to below (or below to above) the centre line, the current reverses direction.

Frequency = the number of times the current (or potential difference) reverses direction in 1 second.

wave trace A

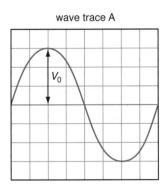

wave trace B

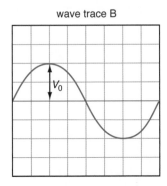

Calculating time period and frequency

Each horizontal division on an oscilloscope screen represents a length of time.

The **time period** is the time to complete one wave (or cycle). If, for trace A, each horizontal division represents 0.005 seconds, then:

time period = 8 divisions × 0.005 = 0.04 s

$$\text{frequency} = \frac{1}{\text{time period}} = \frac{1}{0.04} = 25 \, \text{Hz}$$

examiner tip

You need to be able use oscilloscope traces to compare peak potential differences and to calculate the frequency of a supply.

Mains electricity supply

The UK mains electricity is an a.c. supply at about 230 V. It has a frequency of 50 Hz — this means the current flows one way, then back again, 50 times each second.

Check your understanding

38 How is an a.c. supply different from a d.c. supply?   *(2 marks)*

39 Name one electricity supply that gives a direct current.   *(1 mark)*

40 The diagram shows the oscilloscope traces for three different electricity supplies. The control settings on each oscilloscope are the same.

A

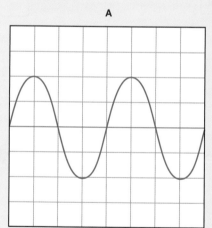

B

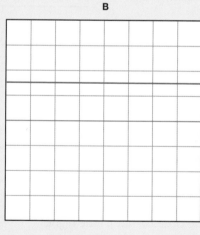

C

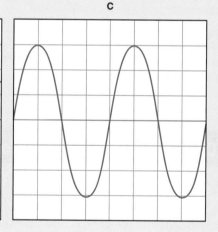

   a) (i)   Which supply, A, B or C, provides a direct current?   *(1 mark)*

   (ii)   The peak potential difference of supply A is 3 V. What is the peak potential difference of supply C?   *(2 marks)*

   b) Each horizontal division on the oscilloscope screen represents 0.002 s.

   Calculate the frequency of trace C.   *(3 marks)*

Answers online   **Test yourself online**   Online

Household electricity

Plugs and cables Revised

Plugs and cables are designed to make appliances safe to use. Looking inside a plug from the back — the b**L**ue wire goes to the **L**eft, the b**R**own wire goes to the **R**ight.

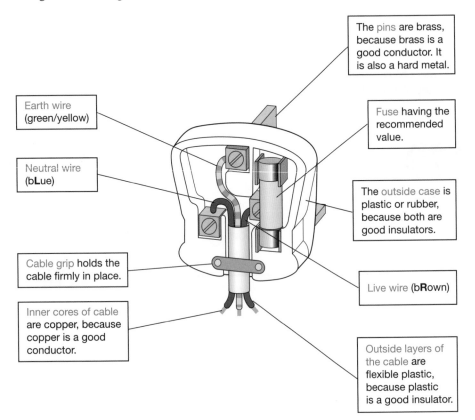

The pins are brass, because brass is a good conductor. It is also a hard metal.

Earth wire (green/yellow)

Fuse having the recommended value.

Neutral wire (b**L**ue)

The outside case is plastic or rubber, because both are good insulators.

Cable grip holds the cable firmly in place.

Live wire (b**R**own)

Inner cores of cable are copper, because copper is a good conductor.

Outside layers of the cable are flexible plastic, because plastic is a good insulator.

← **A correctly wired three-pin plug**

examiner tip
You should be able to spot any mistakes that have been made in wiring a plug and say how to correct them.

Fuses and circuit breakers Revised

● A **fuse** is a thin wire designed to melt and break a circuit when the current exceeds the **current rating** of the fuse.

● The current rating of the fuse should be close to, but a little higher than, the normal current through the appliance.

● A **circuit breaker** is a fast-acting switch that automatically turns a circuit off ('trips') when the current through it goes above a preset value. Once the fault that caused the current to rise is mended, you press a button and the circuit breaker is reset.

● A **residual current circuit breaker** (**RCCB**) automatically switches an appliance off — it detects any difference between the currents in the live and neutral wires of the supply cable. An RCCB operates much faster than a fuse.

Earthed appliances

Revised

Appliances with an outside metal case should be earthed. The **earth wire** (yellow/green) joins the earth pin of the plug to the metal case of the appliance.

● Together, the earth wire and fuse protect the users of electrical appliances from electric shocks.

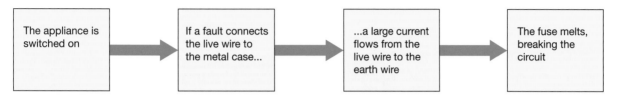

| The appliance is switched on | → | If a fault connects the live wire to the metal case... | → | ...a large current flows from the live wire to the earth wire | → | The fuse melts, breaking the circuit |

● Without the earth wire, the appliance would become **live**. Anyone touching the outside metal case would get a shock as the current flowed through them to earth.

● The earth wire and fuse together also protect the wiring of the circuit from overheating.

← **Appliances carrying this symbol do not need an earth wire — they are double insulated. Any live part is covered in two layers of insulation**

Electricity is dangerous

Revised

Electricity can cause electrocution and fires. Some examples of what you should *not* do include:

● use an adaptor to plug too many appliances into a single socket

● use electrical appliances close to water supplies

● run extension leads into the bathroom

● run cables under carpets

● switch circuits on or off with wet hands

● repair old, frayed cables with tape

> **examiner tip**
> You need to be able to recognise situations when mains electricity is being used dangerously.

Check your understanding

Tested

41 What is the colour of the insulation on the neutral wire inside a cable? *(1 mark)*

42 Why is brass used to make the pins of a plug? *(2 marks)*

43 Explain what happens if the live wire of a cable touches the metal case of an appliance while the appliance is switched on. *(3 marks)*

44 Give one advantage of a circuit breaker over a fuse. *(1 mark)*

Answers online — **Test yourself online** Online

Current, charge, energy and power

Energy and charge

When charge flows through a **resistor**, energy is transferred and the resistor gets hot.

The total energy transferred by a resistor in a circuit depends on:

● the potential difference across the resistor
● how much charge flows through the resistor

The energy transferred is calculated using the equation:

$E = V \times Q$

energy transferred = potential difference × charge

● E in joules, J
● V in volts, V
● Q in coulombs, C

This equation also applies to the electrical energy transferred from a power supply to an appliance.

Power

The rate at which energy is transferred is called **power**. Power is calculated using the equation:

$$P = \frac{E}{t}$$

$$\text{power} = \frac{\text{energy transferred}}{\text{time taken}}$$

● P in watts, W
● E in joules, J
● t in seconds, s

> **examiner tip**
> This equation can be used with any form of energy transfer.

A high-power appliance transfers a lot of energy in a short time. For example, an electric shower with a power rating of 8000 watts (8 kW) transfers 8000 joules of energy every second it is switched on.

Electrical power can also be calculated using the equation:

$P = V \times I$

power = potential difference × current

● P in watts, W
● V in volts, V
● I in amps, A

The information plate on an electrical appliance includes:

● the power of the appliance
● the potential difference of the electricity supply that it requires.

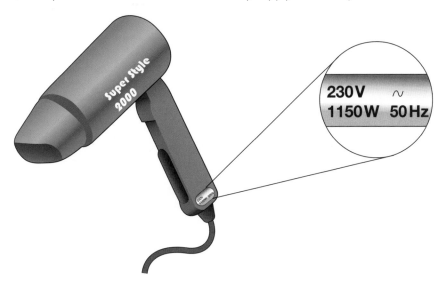

230V ∿
1150W 50Hz

From this you can calculate the current the appliance takes and the size of fuse to use.

examiner tip

To calculate the current — just divide the power by the potential difference. Then choose a fuse with a higher rating than the current.

Check your understanding

Tested

45 a) Calculate the current drawn from the mains by the hair dryer shown in the diagram. *(2 marks)*

b) What size fuse should be fitted to the plug of the hairdryer: 3 A, 5 A or 13 A? *(1 mark)*

46 A 6 V immersion heater transfers 7200 J of energy to a beaker of water in 5 minutes. Calculate:

a) the power of the heater (assume it is 100% efficient) *(2 marks)*

b) the current flowing through the immersion heater *(2 marks)*

c) the total charge flowing through the circuit in 5 minutes *(2 marks)*

47 A 12 V battery drives a car starter motor, taking a current of 90 A for 2 seconds. Calculate:

a) the total charge flow through the motor in 2 seconds *(2 marks)*

b) the electrical energy transferred to the motor *(2 marks)*

48 An electric fan heater has a power of 2.3 kW. It is plugged into the 230 V mains supply and switched on for 2 hours. Calculate:

a) the current drawn by the heater from the mains *(2 marks)*

b) the total charge flow through the heater in 2 hours *(2 marks)*

Answers online —— Test yourself online

Online

Atomic structure

What are atoms made of?

Atoms are made up of protons, neutrons and electrons. The **relative mass** and **relative charge** of protons, neutrons and electrons are given in the table.

Particle	Relative mass	Relative charge
proton	1	+1
neutron	1	0
electron	negligible (about 1/2000)	−1

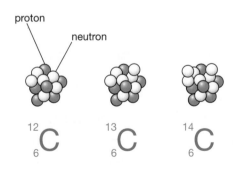

↑ **The basic structure of an atom**

Key
- ⊕ proton
- ○ neutron
- ⊖ electron

For any **atom**:

- number of protons = number of electrons
- the overall charge is zero
- the number of protons is called its **atomic number**
- the total number of protons and neutrons is called its **mass number**.

Example: A copper (Cu) atom is made up of:

     29 protons
     (63 − 29) = 34 neutrons
     29 electrons

$$\text{mass number} = {}^{63}_{29}\text{Cu} = \text{atomic number}$$

An **ion** is an atom that has lost or gained electrons:

- An atom that loses electrons becomes a positive ion.
- An atom that gains electrons becomes a negative ion.

Isotopes

Atoms of different elements have different numbers of protons. All atoms of the same element have the same number of protons.

- **Isotopes** of an element have the same number of protons but different numbers of neutrons.
- Each nucleus in this diagram has six protons.
- So, each is an atom of the same element (carbon).
- But each atom has a different mass number.
- So, each atom has a different number of neutrons.
- The atoms are isotopes of carbon.

proton

neutron

$${}^{12}_{6}\text{C} \quad {}^{13}_{6}\text{C} \quad {}^{14}_{6}\text{C}$$

↑ **The isotopes of carbon**

Scientific models

New evidence may lead to a new scientific model. What scientists knew about atoms at the start of the twentieth century led to the **'plum-pudding' model** of the atom. An atom was thought to be a ball of positively charged material, with negative electrons embedded in it.

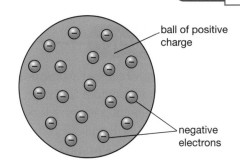

→ Thomson's 'plum-pudding' model of the atom

ball of positive charge

negative electrons

The **alpha-scattering experiment** devised by Ernest Rutherford gave some unexpected results (table below). When alpha particles were fired at a gold foil, instead of going straight though some were deflected through small angles and some rebounded backwards.

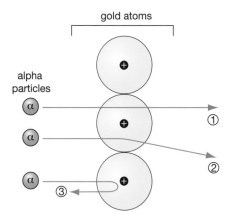

gold atoms

alpha particles

	What happened to the alpha particles	Explanation
Track 1	Most went straight through the foil	Most of the atom is empty space
Track 2	Some were deflected	The nucleus repels the alpha particles so must also be positive
Track 3	A very small number rebounded backwards	The nucleus is very small but has a large mass

The experiment showed that the 'plum-pudding' model was wrong. The **nuclear model** was developed to explain the results.

↑ When positive alpha particles are directed at a very thin sheet of gold foil, they emerge at different angles — most pass straight through the foil, some are deflected and a few appear to rebound from the foil

Check your understanding

49 Which of these nuclei:

  a) have the same mass number?

  b) are isotopes of the same element?

  c) has the largest atomic number? *(3 marks)*

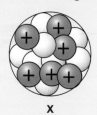
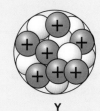
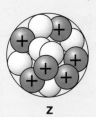

X   Y   Z

50 What happens to change an atom into an ion? *(1 mark)*

51 Which atomic particle was not part of the 'plum-pudding' model of the atom? *(1 mark)*

52 Why did the 'plum-pudding' model of the atom need to be replaced? *(1 mark)*

Answers online   Test yourself online   Online

Atoms and radiation

Unstable nuclei

Atoms of a radioactive substance have unstable nuclei. Radioactive substances emit radiation from the nuclei of their atoms all the time. The rate at which the substance emits radiation (**decays**) does not change, no matter what is done to the substance.

There are three main types of **nuclear radiation**: alpha (α), beta (β) and gamma (γ):

● An **alpha** (α) particle is the same as a **helium nucleus** — 2 neutrons and 2 protons.
● A **beta** (β) particle is an **electron** emitted from the nucleus of an atom.
● **Gamma** (γ) radiation is high-energy **electromagnetic** radiation.

Background radiation

Background radiation gives most people a steady, low **dose** of radiation. Some people receive a higher dose because of:

● where they live
● the type of work they do
● medical treatments

The main sources of background radiation are:

● radon gas — formed when uranium found in various rocks decays
● cosmic rays — constantly bombard the Earth from the stars
● food and drink — small amounts of radioactive materials get into the food chain from the soil
● bricks and other building materials — produced from rocks and clay
● medical sources — mainly X-rays; although not nuclear radiation, X-rays are ionising, and so having an X-ray taken increases your radiation dose

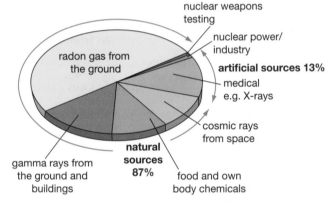

↑ **Sources of background radiation**

Properties of nuclear radiations

Radiation	Ionising power	Range in air	Stopped by	Effect of electric and magnetic fields
Alpha (α)	Strong	A few centimetres	Paper or thin card	Very small deflection
Beta (β)	Moderate	About 1 m	5 mm of aluminium	Large deflection (in the opposite direction to alpha)
Gamma (γ)	Very weak	At least 1 km	About 10 cm lead	None

To explain the effects of **electric fields** and **magnetic fields**:

● Only charged particles are **deflected** — gamma rays are not charged particles.

- Smaller masses are more easily deflected than larger masses — beta particles have a much smaller mass than alpha particles.
- Positive and negative charges are deflected in opposite directions — alpha particles are positive, beta particles are negative.

Radioactive decay

Revised

Atoms that emit an alpha or beta particle change into different elements.

Alpha decay

An atom that emits an alpha particle loses two protons and two neutrons.

Changing the number of protons changes the atom into a different element.

Example: Radium-226 emits an alpha particle and changes into radon-222:

$$^{226}_{88}Ra \rightarrow {}^{222}_{86}Rn + {}^{4}_{2}He$$

> **examiner tip**
>
> Remember that in alpha decay, the top number (mass number) goes down by four; the bottom number (atomic number) goes down by two.

Beta decay

- A neutron in the nucleus splits into a proton and an electron.
- The electron (beta particle) is emitted from the nucleus.
- The proton stays in the nucleus.
- So, an atom that emits a beta particle loses a neutron but gains a proton.

Example: Carbon-14 emits a beta particle and changes into nitrogen-14:

$$^{14}_{6}C \rightarrow {}^{14}_{7}N + {}^{0}_{-1}e$$

> **examiner tip**
>
> Remember that in beta decay, the top number (mass number) stays the same; the bottom number (atomic number) goes up by one.

Check your understanding

Tested

53 In what way are two isotopes of an element the same, and in what way are they different? *(2 marks)*

54 A person flies every week from London to New York and back. Suggest why this increases their background radiation dose. *(2 marks)*

55 An atom of americium-241 ($^{241}_{95}Am$) decays to form an atom of neptunium ($^{237}_{93}Np$).

What type of particle is emitted by americium-241? Explain the reason for your answer. *(2 marks)*

56 Potassium-40 ($^{40}_{19}K$) decays by losing a beta particle to form an isotope of calcium.

a) What is the mass number of the calcium isotope produced? *(1 mark)*

b) What is the atomic number of the calcium isotope? *(1 mark)*

c) Write an equation to show the decay process. *(3 marks)*

Answers online — **Test yourself online** Online

Uses of radioactivity

Radioactive decay is a random process

It is impossible to predict when one particular unstable nucleus will decay. But with a large group of unstable atoms, we can estimate how many nuclei will decay in a period of time. The decay of a large group of unstable atoms always follows the same pattern.

The average number of emissions in a certain time is called the **count rate**.

● The time it takes for the count rate of a radioactive isotope to fall to half its initial value is called its **half-life**. It is also the time it takes for the number of nuclei of the isotope in a sample to halve.

● The shorter the half-life, the faster the isotope decays, and the more unstable it is.

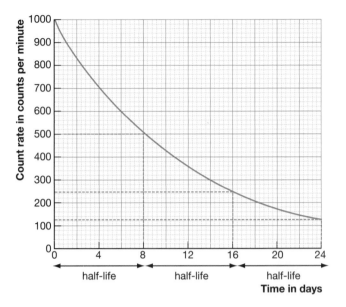

↑ The radioactive isotope iodine-131 has a half-life of 8 days

examiner tip

When you find the half-life from a graph, make sure you look at the units on axes — time could be in seconds, minutes, days or years.

Using isotopes

The use of an isotope depends on its half-life and on the type of radiation emitted. Medical tracers have a short half-life and usually emit gamma radiation. A medical **tracer** is a radioactive isotope attached to a chemical, which is then injected into a patient. The gamma rays are detected outside the patient's body. A short half-life (long enough for doctors to carry out a diagnosis) means that the level of radiation inside the patient's body soon falls to a safe level. Technetium-99 is often used as a medical tracer.

Isotope	Useful radiation emitted	Half-life
Cobalt-60	Gamma	5.3 years
Technetium-99	Gamma	6 hours
Phosphorus-32	Beta	14 days
Manganese-52	Gamma	5.6 days
Strontium-90	Beta	28 years
Radon-220	Alpha	52 seconds

Tracers also have many industrial applications, including:

● locating leaks in underground pipes

● finding the source of river pollution

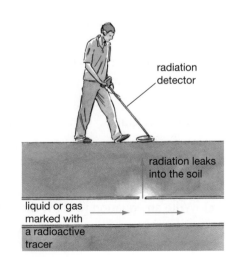

radiation detector

radiation leaks into the soil

liquid or gas marked with a radioactive tracer

Gamma rays kill bacteria, so gamma radiation can be used to:

● **sterilise** medical instruments and dressings

● kill bacteria to keep food fresh for longer (the food does not become radioactive!)

Cobalt-60 is often used to kill bacteria. The long half-life means that the source does not need changing too often and the level of radiation does not change too quickly.

Isotopes used for quality control often have a long half-life. A thick sheet of material will absorb more radiation than a thin sheet of material. So measuring the level of radiation passing through the sheet will monitor its thickness as the sheet is being produced. The control unit responds to the radiation detected, automatically adjusting the pressure the rollers exert on the sheet.

● For materials such as paper, cardboard and plastic, a beta source is used.

● For a metal sheet, a gamma source is used as it has greater penetrating power.

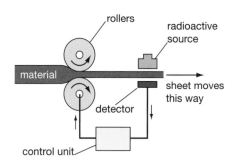

Using nuclear radiation can be dangerous — Revised

Because nuclear radiation can ionise atoms, it is dangerous to living cells.

	Least dangerous	Most dangerous
Outside the body	Alpha — are easily absorbed by the air or by your skin	Gamma and beta — can penetrate the body to reach vital organs
Inside the body	Gamma and beta — can pass through cells without being absorbed	Alpha — strongly absorbed by cells and highly ionising

Ways of reducing exposure to nuclear radiation include:

● shields of lead, concrete or thick glass to absorb the radiation

● reducing the time of exposure

● wearing protective clothing

● handling radioactive materials remotely

Check your understanding — Tested

57 Look at the list of isotopes in the table on page 62. Which isotope would be used in the following situations? Explain the reason for each choice.

a) To monitor the thickness of polythene as it is produced. *(2 marks)*

b) To kill the cells in a cancerous tumour. *(2 marks)*

c) To trace the path of oil in an underground pipe. *(2 marks)*

58 A radioactive isotope contains 128 000 unstable nuclei. After 16 days, this number has gone down to 8000. Calculate the half-life of the isotope. *(2 marks)*

59 What is the number of radioactive emissions from a source in a given time is called? *(1 mark)*

60 Explain why it is important that people are not exposed to high doses of nuclear radiation. *(2 marks)*

> **examiner tip**
> You must consider both the type of radiation emitted and the half-life.

Answers online — Test yourself online — Online

Fission and fusion

Nuclear fission

Nuclear fission is the splitting of an atomic nucleus. **Nuclear reactors** use **nuclear fission** reactions to release large amounts of energy.

The two fuels in common use in nuclear reactors are uranium-235 and plutonium-239 (the majority of reactors use uranium-235).

For fission to happen, a uranium-235 or plutonium-239 nucleus must first capture and absorb a neutron. The new nucleus then splits into two smaller nuclei, plus two or three neutrons. And, very importantly, energy is released.

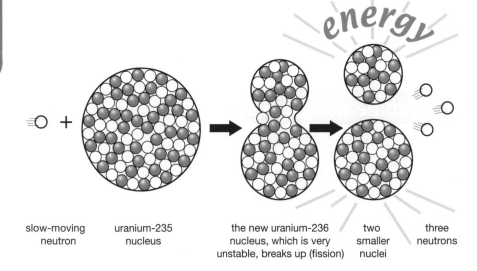

← Fission of a nucleus of uranium-235

slow-moving neutron   uranium-235 nucleus   the new uranium-236 nucleus, which is very unstable, breaks up (fission)   two smaller nuclei   three neutrons

Chain reactions

Every time a fission reaction happens, neutrons are released. These neutrons may go on to cause more uranium or plutonium nuclei to split, releasing even more neutrons. This is called a **chain reaction**.

In a nuclear power station, the chain reaction is controlled in such a way that only one neutron goes on to produce a further fission reaction. This gives a steady, controlled release of energy.

> **examiner tip**
> You must be able to draw a labelled diagram to show how a chain reaction happens.

Nuclear fusion

Nuclear fusion is the joining of two atomic nuclei. During **nuclear fusion** two small nuclei are fused (joined together) to form just one larger nucleus. Each time a fusion reaction happens energy is released.

The nucleus of an atom is positive. So before two nuclei can join together, the repulsion force between the nuclei must be overcome. This only happens at very high temperatures — typically the temperatures found at the centre of a star.

> **examiner tip**
> Make sure you can spell the words 'fusion' and 'fission' and that you know the difference between the two processes.

Nuclear fusion is the process by which energy is released in stars. One of the major fusion processes in stars involves hydrogen nuclei (protons) fusing to form helium. Other fusion processes form larger nuclei such as carbon and oxygen.

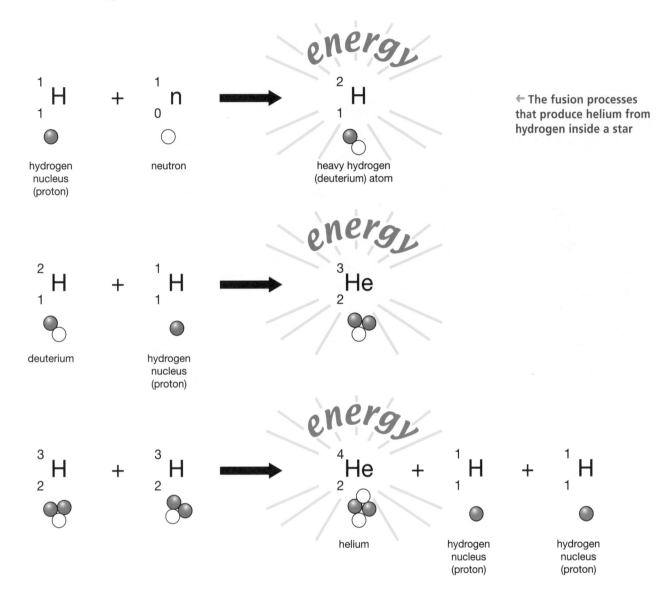

← The fusion processes that produce helium from hydrogen inside a star

61 What must happen to the nucleus of a plutonium-239 atom before fission can happen? *(1 mark)*

62 In a nuclear reactor, control rods are used to absorb neutrons. What effect does this have on the chain reaction? *(1 mark)*

63 How could you tell from the energy output of a nuclear reactor that the chain reaction was speeding up? *(1 mark)*

64 Why is it easier for fusion reactions to happen if the gases are highly compressed, as in the core of a star? *(1 mark)*

Answers online — Test yourself online — Online

Life cycle of a star

Formation of a star

Stars form from a **nebula** — a huge cloud of gas and dust. The gas is mainly **hydrogen**.

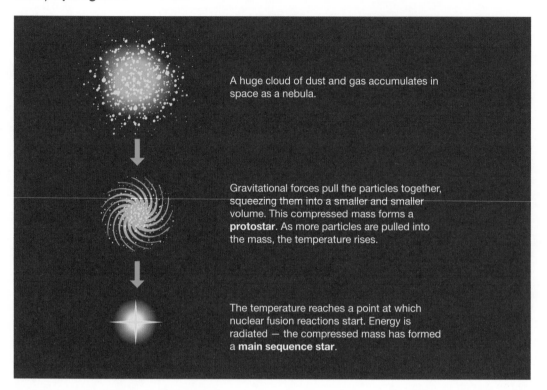

A huge cloud of dust and gas accumulates in space as a nebula.

Gravitational forces pull the particles together, squeezing them into a smaller and smaller volume. This compressed mass forms a **protostar**. As more particles are pulled into the mass, the temperature rises.

The temperature reaches a point at which nuclear fusion reactions start. Energy is radiated — the compressed mass has formed a **main sequence star**.

↑ **The formation of a star from a nebula**

Other masses, too small to form stars, join together to form planets.

Main sequence period

During the **main sequence** period the star is stable. The forces within the star are balanced.

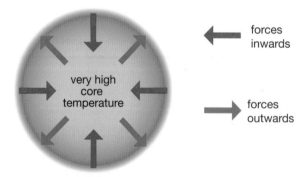

very high core temperature

⟵ forces inwards

⟶ forces outwards

End of the life cycle

What happens to a star once its supply of hydrogen starts to run out depends on the initial mass of the star.

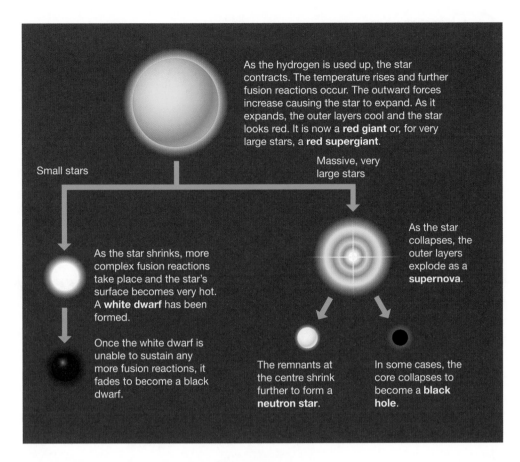

As the hydrogen is used up, the star contracts. The temperature rises and further fusion reactions occur. The outward forces increase causing the star to expand. As it expands, the outer layers cool and the star looks red. It is now a **red giant** or, for very large stars, a **red supergiant**.

Small stars

Massive, very large stars

As the star shrinks, more complex fusion reactions take place and the star's surface becomes very hot. A **white dwarf** has been formed.

Once the white dwarf is unable to sustain any more fusion reactions, it fades to become a black dwarf.

As the star collapses, the outer layers explode as a **supernova**.

The remnants at the centre shrink further to form a **neutron star**.

In some cases, the core collapses to become a **black hole**.

↑ **How a star 'dies' depends on its initial mass**

examiner tip

Make sure that you use the correct terms for each sequence. Saying that the Sun will become a supernova or black hole is wrong.

Nuclear fusion and the elements
Revised

The early universe was mainly hydrogen, but now it contains a variety of elements.

● The elements up to iron were formed by nuclear fusion reactions within main sequence stars.

● Beyond iron, the creation of the heaviest elements occurs only in a supernova.

● A supernova scatters the elements throughout the universe.

Check your understanding
Tested

65 What is a protostar? *(2 marks)*

66 Where are the elements heavier than iron formed? *(1 mark)*

67 Describe what happens to stars like the Sun after the main sequence stage. *(3 marks)*

68 Why will the Sun not change into a red supergiant? *(1 mark)*

Answers online **Test yourself online**
Online

X-rays and CT scans

X-rays are **electromagnetic waves**, with a wavelength similar in size to the diameter of an atom.

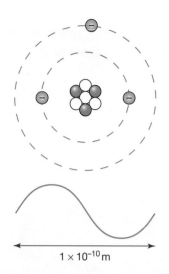

$1 \times 10^{-10}\,\text{m}$

X-ray photography

Revised

X-rays pass through low density materials such as healthy body tissue. Denser materials, such as bone and metals, absorb X-rays.

Just like visible light, X-rays also affect photographic film. An X-ray photograph or 'scan' shows an image of an object having absorbed the X-rays. This makes X-rays useful in the diagnosis of some medical conditions, including bone fractures and dental problems.

→ **The metals pins and bone stand out because they absorb the X-rays and stop them affecting the photograph film**

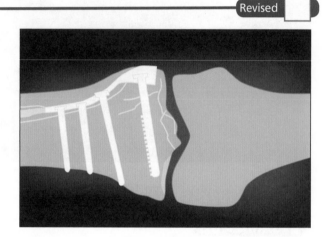

Forming an image electronically

Revised

X-ray images can be formed electronically and displayed on a computer screen. A **charge-coupled device** (**CCD**) is used to convert the transmitted X-rays into electrical charges. A microprocessor then produces the image that is displayed on the screen.

Treatment of cancers

Revised

X-rays are used to kill cancer cells, slow down the growth of a cancer and shrink tumours.

- Low-energy X-rays destroy cancer cells on or near the surface of the body.
- High-energy X-rays, which are more penetrating, are used to deliver radiation to deeper tissue and body organs.

CT scanning

Revised

A **CT** (**computerised tomography**) scanner emits several beams of X-rays from different directions at the same time. The strength of each X-ray beam is measured after it has passed through the patient's body. A computer uses this information to produce a two-dimensional (2D) image showing a slice through the patient's body. A 3D image is produced by joining lots of 'slices' together. In this way, the size and exact position of something, for example a tumour, inside the body can be found.

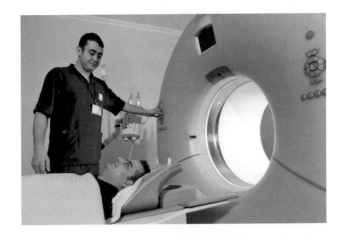

Advantages of CT scans

The images produced:

- are more detailed than ordinary X-ray photographs
- show different types of tissue such as bone, muscle and blood vessels
- can be three-dimensional
- can pinpoint the position of a tumour

Risks associated with X-rays

Revised

X-rays are **ionising** radiations. They can be absorbed by cells and will ionise some atoms, which may cause changes to the molecules that control the way a cell works. This may cause gene **mutation**, damage to the central nervous system or **cancer**.

A CT scan delivers a higher dose of radiation to the body than a normal X-ray. But the increased risk to health from a CT scan must be considered against the benefit of being able to diagnose a problem more accurately.

examiner tip

Given relevant data, you need to be able to compare the relative risks to health from X-rays and CT scans.

Reduce the risk of cell damage by reducing exposure

Our bodies recover quickly from small doses of X-rays. So having a few X-rays taken is not hazardous. But people who operate X-ray equipment as part of their daily job could be **exposed** to large doses, so often they work behind concrete, lead or thick glass screens. These materials reduce **exposure** by absorbing stray X-rays.

A radiographer operating a CT scanner would normally do so from outside the room. If a child is having a CT scan, a parent may be allowed to stay inside the room. The parent would then wear a lead apron to minimise exposure to any stray X-rays.

Check your understanding

Tested

1 Why should metal jewellery be removed before an X-ray is taken?
(1 mark)

2 Why is X-ray film kept in lightproof containers? (1 mark)

3 Why are X-rays dangerous? (2 marks)

4 Some CT scanners are able to scan a section of the body in only a few seconds. What is the advantage to the patient of a CT scan being produced quickly? (1 mark)

5 What precautions are taken by radiographers to reduce their exposure to stray X-rays? (1 mark)

Answers online — **Test yourself online** Online

Ultrasound

What is ultrasound?

The **range** of normal hearing is from about 20 Hz up to 20 000 Hz.

Ultrasound:

● is any sound with a frequency too high for humans to hear

● is any sound with a frequency above 20 000 Hz

● can be produced electronically

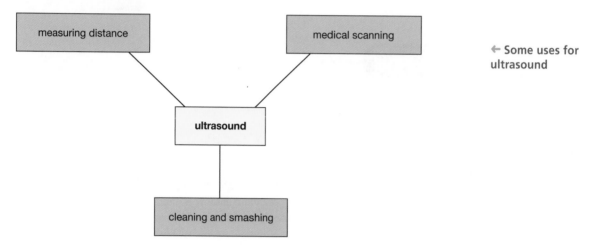

← Some uses for ultrasound

Measuring distance with ultrasound

When ultrasound meets a boundary between two different materials, some will be reflected and some will be transmitted.

A detector can pick up the pulse of reflected ultrasound and display it on an oscilloscope screen. The longer it takes the reflected pulse to reach the detector, the further the boundary is from the detector.

Moving an ultrasound transmitter across a surface produces multiple reflections. A computer uses the reflected pulses to form an image.

Remember — each horizontal division on an oscilloscope represents a certain length of time. This is the **time base** setting.

Example

Time base = 2 ms/cm

Each centimetre across the oscilloscope screen represents 2 ms (0.002 s).

If the speed of the ultrasound is known, distances can be calculated using the equation:

$s = v \times t$

distance = speed × time

● s in metres, m

● v in metres per second, m/s

● t in seconds, s

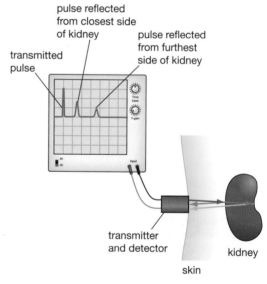

↑ A simplified diagram of the oscilloscope trace produced when an ultrasound pulse is transmitted through a kidney — the settings on the oscilloscope can be adjusted to measure the size of the kidney

examiner tip

In the time between the pulse being transmitted and the reflected pulse being detected, the ultrasound has travelled to a boundary *and* back.

Medical scanning

Pulses of ultrasound are transmitted directly through the mother to the fetus.

● Ultrasound is partially reflected every time it meets a boundary between different tissue types.

● A detector picks up the reflected pulses, which are processed by a computer to give an image of the fetus.

← The combined ultrasound transmitter and detector is moved over the skin on a layer of gel — without the gel, most of the ultrasound would be reflected at the air–skin boundary

Ultrasound scans are also used to investigate other areas of the body, for example:

● damaged ligaments

● the stomach, liver, heart and kidneys

● tumours

Cleaning and smashing

Ultrasound vibrations can be used:

● to remove tartar from teeth

● to smash kidney stones without the need for an operation

Comparing ultrasound and X-ray imaging
Revised

Ultrasound	X-rays
Non-ionising — will not damage body cells; no special precautions needed	Ionising radiation — can damage or kill healthy body cells; operators must protect themselves
Scans can be produced over a long period of time	Harmful in large doses; used only for short amounts of time
Images are less detailed than X-rays	Images are more detailed than ultrasound

examiner tip

In comparing ultrasound with X-rays, the most important thing to remember is that ultrasound is non-ionising and X-rays are ionising.

Check your understanding
Tested

6 Why must gel be used between the ultrasound transmitter and the skin of a woman having a pre-natal scan? *(1 mark)*

7 Ultrasound travels through bone at about 4000 m/s. Calculate the time taken for ultrasound to travel across a bone 2 cm thick. *(2 marks)*

8 Give one advantage and one disadvantage of an ultrasound scan compared to an X-ray. *(2 marks)*

9 Explain why ultrasound is used for pre-natal scanning and not X-rays. *(2 marks)*

Answers online — **Test yourself online**
Online

Lenses

Refraction

Revised

When a ray of light crosses the boundary between two different transparent materials, it usually changes direction. This change in direction is called **refraction**.

A ray of light travelling into a more dense medium, such as glass, from a less dense medium, such as air, always refracts towards the normal. This means that the angle of incidence (*i*) in the air is bigger than the angle of refraction (*r*) in the glass.

The angle of incidence and the angle of refraction are linked by the equation:

$$\text{refractive index} = \frac{\sin i}{\sin r}$$

The higher the **refractive index** of a medium, the more the light changes direction when it enters that medium from air.

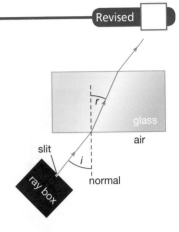

> **examiner tip**
>
> A refractive index is always bigger than 1, so always divide the smaller number into the bigger number. Refractive index does not have a unit.

> **examiner tip**
>
> Refraction and reflection — similar sounding words but totally different processes — do not confuse them.

↑ **Light is not refracted when it travels at 90° to a boundary — i.e. along the normal**

Converging lenses

Revised

Converging lenses (also called convex lenses) are thicker in the middle than at the edges.

Parallel rays of light refracted by a converging lens pass through the **principal focus**. The distance between the principal focus and the lens is called the **focal length** of the lens.

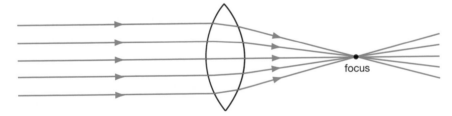

focus

← **There is a principal focus on each side of the lens, the same distance away**

Image formed by a converging lens

← **When the image is seen on a piece of paper (screen), the image must be real**

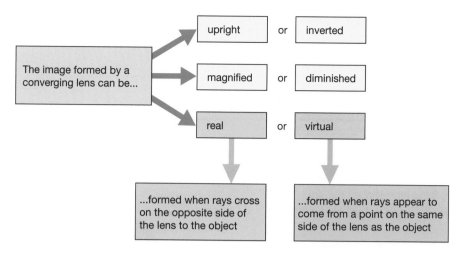

The position and nature of the image formed by a lens is found by drawing a **ray diagram**.

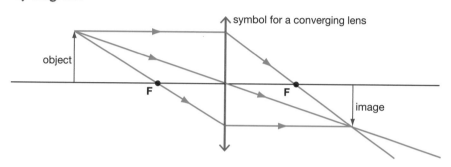

← The image produced when the object is a long way from a converging lens is real, diminished and inverted

examiner tip
This diagram shows three construction lines — to find the position of an image you draw any two of them.

Using a converging lens as a magnifying glass

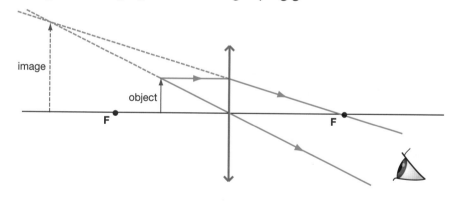

← When the object is between the lens and the principal focus, the image is virtual, upright and magnified

Magnification
Revised

The **magnification** produced by a lens is calculated using the equation:

$$\text{magnification} = \frac{\text{image height}}{\text{object height}}$$

Magnification	Size of image
Less than 1	Smaller than object
Equal to 1	Same as object
Bigger than 1	Bigger than object

- The object height and the image height must have the same unit.
- Magnification does not have a unit.
- Magnification is always a positive number.

Diverging lenses
Revised

Diverging lenses (also called concave lenses) are thinner in the middle than at the edges.

Parallel rays of light refracted by a diverging lens spread out as if they come from a single point — this point is the **principal focus**. It is a **virtual focus**.

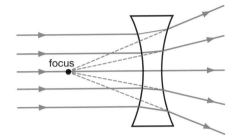

Image formed by a diverging lens

Ray diagrams are drawn for diverging lenses in the same way as those for converging lenses.

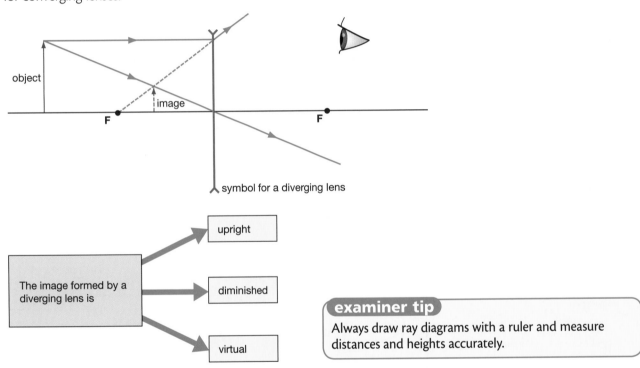

symbol for a diverging lens

```
The image formed by a        →    upright
diverging lens is            →    diminished
                             →    virtual
```

examiner tip

Always draw ray diagrams with a ruler and measure distances and heights accurately.

Check your understanding
Tested

10 Light entering a glass block at an angle of 50° to the normal is refracted at an angle of 30°. Calculate the refractive index of the glass. *(2 marks)*

11 Describe the difference between a converging lens and a diverging lens. *(2 marks)*

12 An object is placed 4 cm in front of a converging lens of focal length 6 cm. Describe the nature of the image formed by the lens. *(2 marks)*

13 An object is placed 10 cm in front of a converging lens of focal length 5 cm. Draw a ray diagram to find the position and nature of the image. *(4 marks)*

14 The image formed using a diverging lens is 4 cm tall. The object is 16 cm tall. Calculate the magnification produced by the lens. *(2 marks)*

Answers online — Test yourself online ⟩ Online

Structure of the eye and camera

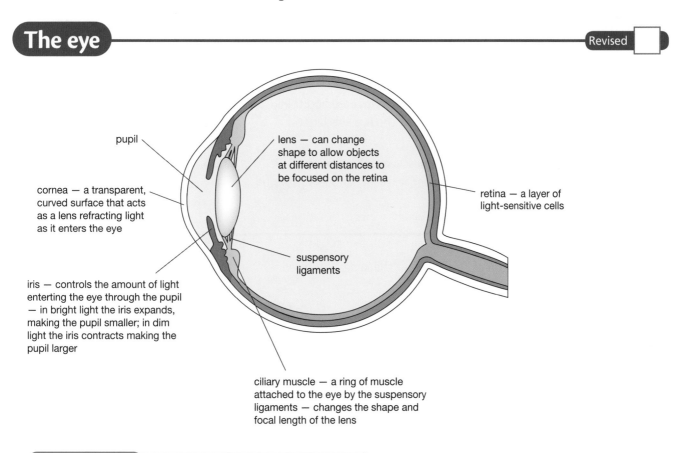

pupil

lens — can change shape to allow objects at different distances to be focused on the retina

cornea — a transparent, curved surface that acts as a lens refracting light as it enters the eye

retina — a layer of light-sensitive cells

suspensory ligaments

iris — controls the amount of light enterting the eye through the pupil — in bright light the iris expands, making the pupil smaller; in dim light the iris contracts making the pupil larger

ciliary muscle — a ring of muscle attached to the eye by the suspensory ligaments — changes the shape and focal length of the lens

examiner tip

Learn the names and functions of the parts of the eye shown in the diagram.

The camera

A camera and an eye have many similar features. Both have:

- a lens
- a way of changing the amount of light entering
- a light-sensitive surface

The **image** produced by a camera is always real and inverted.

In a camera the **aperture** is equivalent to the pupil. The size of the aperture is controlled by the **diaphragm**, equivalent to the **iris**.

In a traditional camera, the **photographic film** is equivalent to the **retina** of the eye. In a **digital camera** the photographic film is replaced by a **charge-coupled device** (CCD).

Check your understanding

15 Which part, or parts, of the eye focus light onto the retina? *(1 mark)*

16 What is the function of the iris and pupil? *(2 marks)*

17 How is the lens in a camera similar to the lens in an eye? *(1 mark)*

Answers online — **Test yourself online**

Correcting vision

Range of vision

A normal eye can focus on any object that is between about 25 cm from the eye (the **near point**) and infinity (the **far point**).

● To focus on near objects, the ciliary muscle is contracted and the suspensory ligaments are slack — this makes the eye lens fat and powerful.

● To focus on distant objects, the ciliary muscle is relaxed — the suspensory ligaments pull the lens, making it thinner and less powerful.

Long sight

The eye cannot focus clearly on close objects — the image is formed behind the retina.

Long sight can be caused by:

● the eyeball being too short

● the cornea being too flat

● the eye lens being too weak

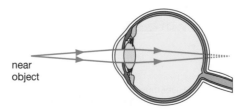

near object

Correcting long sight

A converging lens effectively increases the power of the eye lens and cornea. Close objects now focus on the retina.

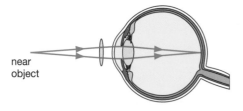

near object

Short sight

The eye cannot focus clearly on distant objects — the image is formed in front of the retina.

Short sight can be caused by:

● the eyeball being too long

● the cornea being too curved

● the eye lens unable to focus

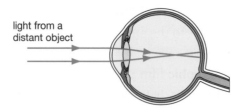

light from a distant object

Correcting short sight

A diverging lens is used to make the light spread out before entering the eye. Distant objects now focus on the retina.

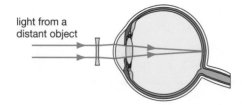

light from a distant object

Power of a lens

The **power** of a lens is determined by its **focal length** — the shorter the focal length, the stronger the power of the lens.

Power and focal length are linked by the equation:

$$P = \frac{1}{f}$$

$$\text{power} = \frac{1}{\text{focal length}}$$

● P in dioptres, D

● f in metres, m

Converging lenses have a **positive power**, such as +0.2 D; diverging lenses have a **negative power** such as −0.4 D.

The focal length of a lens depends on:

● the refractive index of the material used to make the lens

● the curvature of the front and back surfaces of the lens

To make a lens of a particular power, it could be made from a variety of different types of glass or plastic, each with a different refractive index. A lens made from a material with a high refractive index will be flatter than a lens of the same power made from a material with a lower refractive index. This means that the lens can be made thinner — useful for making spectacles for example.

Check your understanding

18 What is meant by the 'near point' of the eye? *(1 mark)*

19 What type of lens is used to correct short sight? *(1 mark)*

20 What is the power of a diverging lens of focal length 20 cm? *(2 marks)*

21 A lens has a power of +0.4 D. What type of lens is it and what is its focal length? *(3 marks)*

22 A manufacturer produces spectacle lenses from materials of different refractive index. For a particular power, which material gives the thickest lenses? *(1 mark)*

Answers online **Test yourself online**

Other applications using light

Total internal reflection can only happen when light travels from a dense medium, like glass, into a less dense medium, usually air.

For total internal reflection to happen, the light must hit the boundary with the air at an angle of incidence greater than critical angle.

The **critical angle** is the angle of incidence that gives an angle of refraction of 90°.

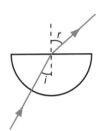

angle of incidence (*i*) < critical angle (*c*)
so that the light is refracted into the air

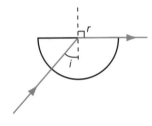

angle of incidence (*i*) = critical angle (*c*)
angle of refraction (*r*) = 90°

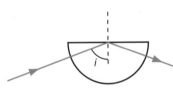

angle of incidence (*i*) > critical angle (*c*)
so that the light is totally internally reflected

Refractive index and critical angle are linked by the equation:

$$\text{refractive index} = \frac{1}{\sin c}$$

examiner tip

This is one of the few equations in which there are no units.

A simple optical fibre is a thin tube of glass. Light rays going in at one end are totally internally reflected along the fibre until they leave at the other end.

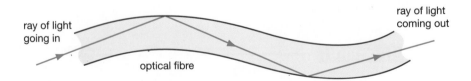

ray of light
going in

optical fibre

ray of light
coming out

Medical uses

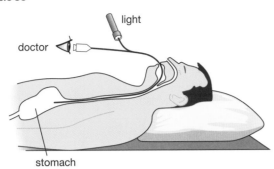

An **endoscope** is used by a doctor to look directly inside a patient's body.

● An endoscope has two bundles of optical fibres.

● Light is sent down one bundle, and reflected back through the second bundle.

● The image is looked at through a lens — or if a miniature CCD camera is used the image can be projected onto a computer screen.

An endoscope may also have a channel through which miniature surgical instruments can be passed.

examiner tip

You need to be able to explain how an endoscope works.

Lasers in medicine

Revised

Lasers have a growing number of medical applications.

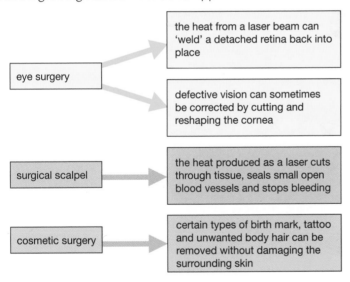

Check your understanding

Tested

23 The critical angle for diamond is 24.4°. Calculate the refractive index of diamond. *(2 marks)*

24 Give the two conditions necessary for total internal reflection to happen. *(2 marks)*

25 Describe the basic principles of an endoscope. *(2 marks)*

26 Give one advantage of using a laser to cut through human tissue rather than using a traditional scalpel. *(1 mark)*

Answers online — **Test yourself online**

Online

Centre of mass and stability

The **centre of mass** of an object is the point at which we can think of the whole mass of the object to be concentrated.

If an object is symmetrical, the centre of mass is always on an **axis of symmetry**.

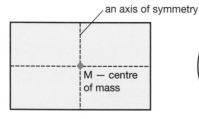
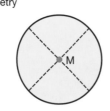
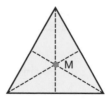

When an object is suspended, it comes to rest with its centre of mass directly below the point of suspension.

← The centre of mass does not have to be inside the object itself

---

Finding the centre of mass Revised

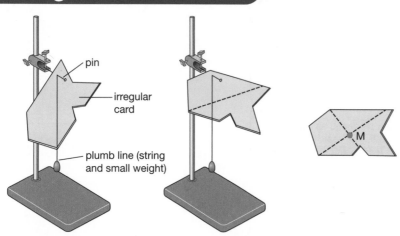

pin

irregular card

plumb line (string and small weight)

1 Use a long pin to suspend the card from one point.

2 Attach a plumb-line to the pin.

3 Mark the position of the plumb-line on the card.

4 Now suspend the card and plumb-line from a different point.

5 Draw a second vertical line on the card.

6 Where the two lines cross is the centre of mass of the card.

> **examiner tip**
>
> Learn these steps — you must be able to describe how to find centre of mass of an irregular sheet of card.

Simple pendulum

A simple pendulum is a small mass (called 'the bob') tied to a piece of string, which hangs from a support. Pulling the bob a small distance to one side and then letting go makes the pendulum swing backwards and forwards.

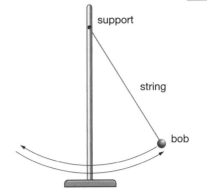

- The **time period** is how long it takes for the pendulum to swing from one side to the other, and back again.
- The **frequency** of a pendulum is how many complete swings it makes in one second.

The time period and frequency of a pendulum are linked by the equation:

$$T = \frac{1}{f}$$

$$\text{time period} = \frac{1}{\text{frequency}}$$

- T in seconds, s
- f in hertz, Hz

The time period of a pendulum depends on the length of the pendulum — the longer the pendulum, the longer the time period.

Stability

Many objects are designed to be stable. This means that if they are tilted slightly and then released they do not fall over.

Objects designed to be stable have:

- a low centre of mass
- a large base area

→ The design of a racing car makes it a stable vehicle

Check your understanding

27 Where is the centre of mass of a car tyre? *(1 mark)*

28 A pendulum makes 10 complete swings every 5 seconds. What is the frequency and time period of the pendulum? *(2 marks)*

29 Explain how a Bunsen burner has been designed to be stable. *(2 marks)*

30 Which one of these three beer glasses would be easiest to knock over? Explain the reason for your answer. *(2 marks)*

Answers online   Test yourself online

Moments

A force that acts on a object but does not go though the pivot will have a turning effect:

● the **pivot** is a point about which a body can rotate
● the **turning effect** of a force is called the **moment**

pivot

force

← The force applied to the spanner has a turning effect on the nut — the longer the spanner, the larger the turning effect

The size of a moment is calculated using the equation:

$$M = F \times d$$

moment = force × perpendicular distance from the line of action of the force to the pivot

● M in newton metres, N m
● F in newtons, N
● d in metres, m

line of action of force

pivot

perpendicular distance

↑ Perpendicular means 'at right angles' — so in the equation use the shortest distance between the pivot and the line along which the force acts

> **examiner tip**
>
> If you forget the unit, just look at the equation — it's the unit of force (N) multiplied by the unit of distance (m).

> **examiner tip**
>
> In any situation, if you need to explain how to increase the moment of a force without increasing the force itself, look for a way of increasing the perpendicular distance. For example, using a longer spanner.

Levers
Revised

A **lever** is a simple machine that helps us to do a job. We apply a force to the lever, which then produces a larger force on an object. The lever increases the force applied to it — the lever is a **force multiplier**.

→ A force applied to the handle of the hammer produces a larger force on the nail — the hammer is a force multiplier

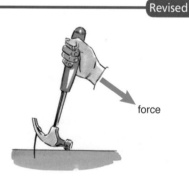

force

Balancing moments
Revised

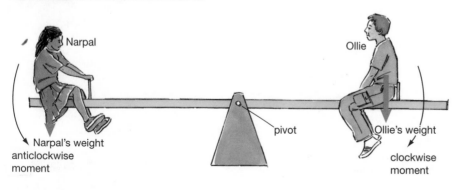

Narpal

Ollie

pivot

Narpal's weight
anticlockwise
moment

Ollie's weight

clockwise
moment

← The see-saw is in equilibrium — it is balanced and is not turning. The clockwise moment caused by Ollie and the anticlockwise moment caused by Narpal must be equal.

The **principle of moments** states that if an object is not turning:

total **clockwise moment** about the pivot	**=**	total anti**clockwise moment** about the pivot

Using the data in the diagram you should be able to calculate and show that:

moment caused by weight of the ladder	**=**	moment caused by weight of the bucket of water

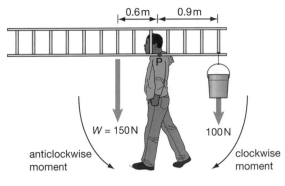

↑ **The ladder is not turning**

In a moments question, where a force or a distance is unknown you should be able to use the principle of moments to calculate it.

examiner tip

Remember that when more than one force acts on one side of the pivot, each force will cause a moment and needs including in the calculation.

Moments and stability
Revised

A vertical line down from the centre of mass of an object shows the line of action of the weight of the object. If the line of action falls outside the object's base, the object will start to topple over.

The moment caused by the lamp's weight (W) makes the lamp fall back to its original position — the lamp is **stable**

The moment caused by the lamp's weight (W) makes the lamp topple over — the lamp is **unstable**

Making objects more stable is about lowering the centre of mass and increasing the base area. This allows the object to be tilted through a bigger angle before the line of action of the weight falls outside the base.

Check your understanding
Tested

31 A spanner, 30 cm long, is used to undo a nut. A force of 85 N is applied to the end of the spanner. Calculate, in N m, the moment exerted on the nut. *(2 marks)*

32 What is meant by a lever being a 'force multiplier'? *(1 mark)*

33 Narpal weighs 360 N and sits 2 m from the middle of a see-saw. Ollie weighs 480 N. How far from the middle of the see-saw should Ollie sit in order for the see-saw to balance? *(3 marks)*

34 Describe how the design of the table lamp shown in the diagram above could be changed in order to make it more stable. *(2 marks)*

Answers online ——— **Test yourself online** Online

Hydraulics

Pressure

Revised

Pressure is caused when a force acts on an area. Pressure is calculated using the equation:

$$P = \frac{F}{A}$$

$$\text{pressure} = \frac{\text{force}}{\text{area}}$$

- P in pascals, Pa (N/m²)
- F in newtons, N
- A in metres squared, m²

Pressure in a liquid

Revised

The liquid inside a container causes a pressure on all sides of the container.

→ **At the same depth, the pressure in a liquid acts equally in all directions**

water

Hydraulic machines

Revised

Hydraulic machines use a liquid to transmit pressure. This makes it possible for a hydraulic machine to transmit and magnify a force.

To be able to do this, a hydraulic machine makes use of two properties of liquids:

- Liquids are virtually **incompressible** — the volume hardly changes no matter how hard the liquid is squeezed.
- Pressure is transmitted equally in all directions throughout a liquid.

Hydraulic jacks

A hydraulic jack is used to lift a heavy weight (**load**) using a small force (**effort**).

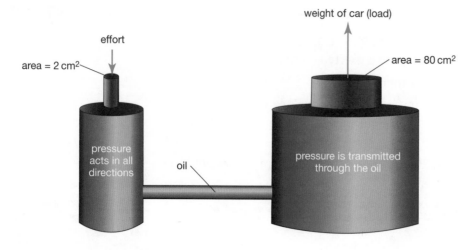

weight of car (load)

effort

area = 2 cm²

area = 80 cm²

pressure acts in all directions

oil

pressure is transmitted through the oil

This, like many other hydraulic machines, is acting as a **force multiplier**. This only happens when the area of the piston to which the effort is applied is smaller than the area of the piston acting on the load.

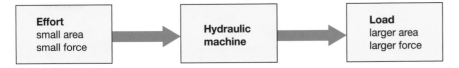

| Effort
small area
small force | → | Hydraulic
machine | → | Load
larger area
larger force |

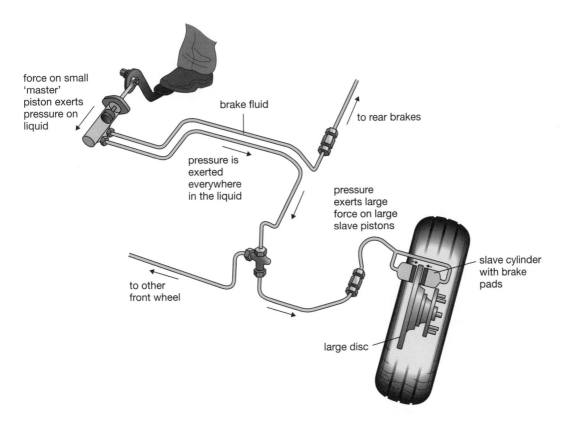

force on small 'master' piston exerts pressure on liquid

brake fluid

to rear brakes

pressure is exerted everywhere in the liquid

pressure exerts large force on large slave pistons

slave cylinder with brake pads

to other front wheel

large disc

The brake fluid transmits the pressure equally in all directions — this makes the pressure at each wheel the same. If the slave pistons each have the same area, the braking force at each wheel will also be the same.

Check your understanding   Tested

35 A skier weighs 768 N. Each ski has an area of 0.192 m². Calculate the pressure on the snow when the skier has both skis flat on the ground.   (2 marks)

36 What is meant by a liquid being 'virtually incompressible'?   (1 mark)

37 Explain the advantage of a mountain bike having wide rather than narrow tyres   (2 marks)

38 In a hydraulic brake system, which has the largest cross-sectional area — the 'master' piston or one of the 'slave' pistons?   (1 mark)

Answers online — **Test yourself online**   Online

Circular motion

Moving in a circle

Remember:

- Velocity includes both speed and direction.
- Acceleration is the rate of change of velocity.

An object moving round in a circle is continually changing direction. This means that its velocity is continually changing, even if its speed stays the same. So, a body moving in a circle must be accelerating:

- Any object moving in a circle is continually accelerating towards the centre of the circle.
- The acceleration changes the direction of the object, but not its speed.

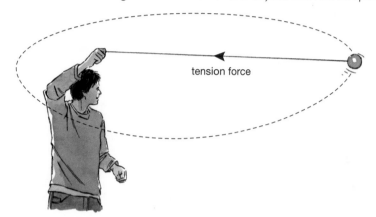

tension force

← The ball is changing direction, so it must be accelerating

For an object to accelerate there must be a resultant force acting on it. The force causing the acceleration of the ball is the inward pull on the ball through the string. This is the **tension force** in the string. Just like the acceleration of the ball, the tension force acts towards the centre of the circle.

examiner tip

If you are asked a question involving an object travelling at a constant speed and accelerating at the same time, the answer is to do with change in direction.

Centripetal force

The force that makes an object move in a circle is called **centripetal force**. This force always acts inwards, towards the centre of the circle.

The centripetal force needed to keep an object moving in a circle increases if:

- the mass of the object increases
- the speed of the object increases
- the radius of the circle decreases

The speed of the object has the biggest effect on the size of the centripetal force — doubling the speed of an object increases the centripetal force four times.

In different situations, different types of force provide the centripetal force.

With the hammer thrower, the **tension** in the wire provides the centripetal force acting on the rotating 'ball'.

tension

The **friction** between the road and the motorbike's tyres provides the centripetal force needed to change direction.

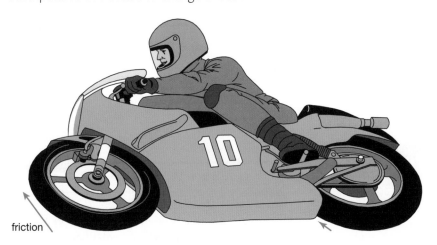

friction

The centripetal force acting on a satellite orbiting the Earth is provided by the **gravitational** pull of the Earth on the satellite.

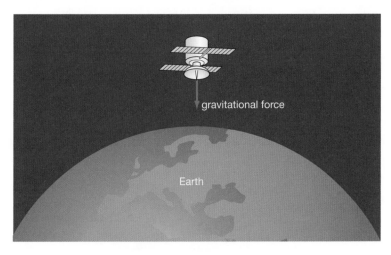

gravitational force

Earth

The Earth is kept in orbit around the Sun by the gravitational pull of the Sun on the Earth.

Check your understanding ———————————————— Tested

39 Explain why a cyclist racing on a circular track is continually accelerating. *(2 marks)*

40 If an object is to accelerate, what must act on the object? *(1 mark)*

41 Tom weighs 300 N and his brother Sam 200 N. They are standing the same distance from the centre of a roundabout ride. The roundabout is rotating quickly.

a) Compare the size of the centripetal force acting on each of the two boys. *(1 mark)*

b) What happens to the size and the direction of the centripetal force acting on the boys as the roundabout slows down? *(2 marks)*

42 Which force provides the centripetal force keeping the Moon in orbit around the Earth? *(1 mark)*

Answers online ——— **Test yourself online** ———————————— Online

The motor effect

Electromagnets

An electric current flowing through a wire produces a **magnetic field**. The magnetic field lines around a straight wire are a series of concentric circles.

Passing an electric current through a wire that is wrapped around a piece of iron (iron **core**) creates an **electromagnet**.

Applications of electromagnets

Relay switch

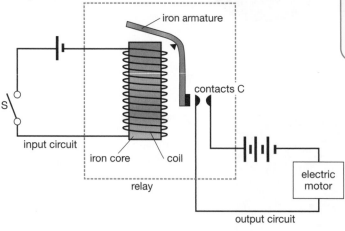

input circuit · iron core · coil · relay · contacts C · iron armature · electric motor · output circuit

> **examiner tip**
>
> If presented with a diagram and asked to explain how a device works, start with what happens when the electromagnet is switched on.

- Closing the switch S completes the input circuit.
- The electromagnet becomes magnetised.
- The iron armature is attracted to the electromagnet.
- As the armature pivots, it closes the contacts C.
- The output circuit is now complete.
- An electric current flows through the motor.

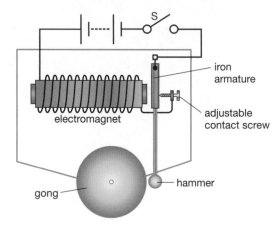

electromagnet · iron armature · adjustable contact screw · hammer · gong

↑ **An electric bell uses an electromagnet**

The motor effect

A current-carrying conductor placed in a magnetic field experiences a force — this is called the **motor effect**.

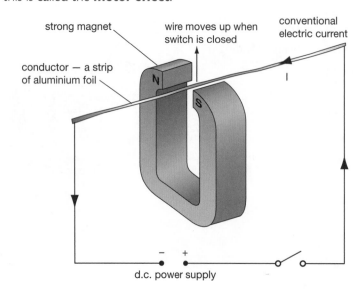

strong magnet · wire moves up when switch is closed · conventional electric current · conductor — a strip of aluminium foil · d.c. power supply

← When the switch is closed, the aluminium strip experiences a force and moves upwards

> **examiner tip**
>
> Do not use the word 'bigger' when you mean 'stronger'. Using a bigger magnet may not increase the force on a conductor, but using a stronger magnet will.

The size of the force on the aluminium strip is increased if:

● the strength of the magnetic field is increased
● the size of the current is increased

Note: the conductor does not experience a force if it is parallel to the magnetic field lines.

The direction of the force can be found using **Fleming's left-hand rule**. Reversing the magnetic field or reversing the current will reverse the direction of the force.

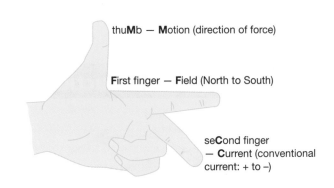

thuMb — **M**otion (direction of force)

First finger — **F**ield (North to South)

se**C**ond finger — **C**urrent (conventional current: + to −)

Simple d.c. electric motor

When a current flows through the coil, the coil rotates. This is due to the motor effect and happens because:

● the forces acting on each side of the coil, X and Y, are in opposite directions
● each force produces a moment, causing the coil to rotate (in this case anticlockwise)
● whenever the coil passes the vertical position, the direction of the current and the direction of forces acting on X and Y reverse
● the coil continues to rotate as long as the current flows

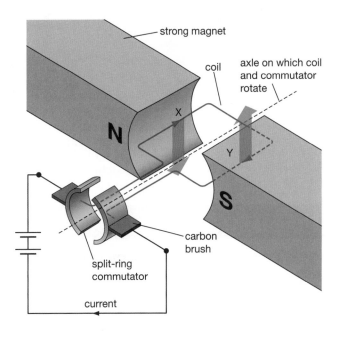

strong magnet

coil

axle on which coil and commutator rotate

carbon brush

split-ring commutator

current

> **examiner tip**
>
> Make sure you can use Fleming's left-hand rule to work out the direction of rotation of the coil in an electric motor diagram.

Loudspeaker

A loudspeaker also works because of the motor effect.

● A current through the movable coil produces a force either forwards or backwards on the coil.
● An alternating current causes the force to swap direction and the coil to vibrate.
● The coil causes the diaphragm and speaker cone to vibrate.
● The vibrations create sound waves that travel through the air to the listener.

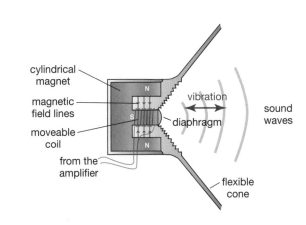

cylindrical magnet

magnetic field lines

moveable coil

from the amplifier

vibration

diaphragm

sound waves

flexible cone

Check your understanding
Tested

43 How is the way that an electric bell works similar to the way that a relay switch works? *(1 mark)*

44 A current passes through a copper wire placed between the poles of a magnet. Explain why the wire does not move. *(2 marks)*

45 Give two ways that the direction of rotation of a motor can be reversed. *(2 marks)*

46 How will increasing the current to a loudspeaker coil change the vibration of the diaphragm? *(1 mark)*

Answers online — **Test yourself online**
Online

Transformers

Electromagnetic induction

Revised

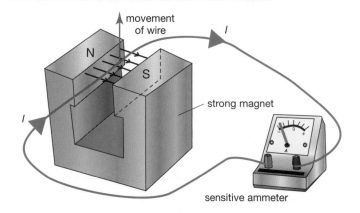

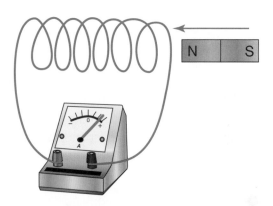

⬆ Moving the wire up or down through the magnetic field causes a potential difference (p.d.) to be induced across the ends of the wire. An induced current flows in the wire

⬆ An induced p.d. is also produced if a bar magnet is pushed into or pulled out of a stationary coil of wire causing an induced current in the coil

How do transformers work?

Revised

A transformer has two coils of **insulated wire** wound on a soft **iron core**.

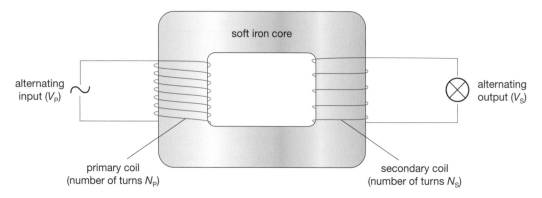

- An alternating current in the primary coil produces an alternating magnetic field that magnetises the iron core.
- The magnetic field lines in the iron core pass through the secondary coil.
- An alternating potential difference is induced across the ends of the secondary coil.
- If the secondary coil is part of a complete circuit a current is induced in the coil.

Transformers are either **step-up** (increase the potential difference) or **step-down** (decrease the potential difference).

Step-up	Step-down
$V_p < V_s$	$V_p > V_s$
$N_p < N_s$	$N_p > N_s$

The potential differences across the primary and secondary coils of a transformer are linked by the equation:

$$\frac{V_P}{V_S} = \frac{N_P}{N_S}$$

- V_p and V_s are both in volts, V
- N_p and N_s are numbers and have no units

> **examiner tip**
>
> If a transformer increases the p.d. it is a step-up transformer. So you should expect more turns on the secondary coil than on the primary coil.

If transformers were 100% efficient, then:

electrical power input = electrical power output

$$V_p \times I_p = V_s \times I_s$$

- V_p and V_s are both in volts, V
- I_p and I_s are both in amps, A

Switch-mode transformers

Revised

Switch-mode transformers are found in such devices as the power supplies for mobile phone chargers. Switch-mode transformers work at a high frequency — typically anything between 50 kHz and 200 kHz. The higher the frequency used, the smaller and lighter the transformer can be made.

The power loss is small, with efficiencies as high as 95%. Even when left switched on with no load connected to the output, there is little power wasted.

The advantages of switch-mode transformers compared to traditional 50 Hz mains transformers are that they are:

- much smaller
- much lighter
- very efficient

Check your understanding

Tested

47 Would a potential difference be induced if a bar magnet is held stationary inside a coil of wire? Give a reason for your answer.

(2 marks)

48 A transformer is used to operate a portable 9 V radio from the 230 V mains electricity supply. If there are 36 turns on the secondary coil, how many turns are there on the primary coil?

(2 marks)

49 A step-down transformer is used to connect a 110 volt, 575 watt electric cement mixer to the 230 volt mains supply. Calculate the current drawn from the mains supply. What assumption did you make in order to complete this calculation? *(3 marks)*

50 How big is a switch-mode transformer compared to a traditional mains transformer? *(1 mark)*

Answers online — Test yourself online — Online

Index